CONSERVATION METHODS FOR Terrestrial Orchids

NIGEL D. SWARTS
KINGSLEY W. DIXON

J.ROSS PUBLISHING

ISBN-13: 978-1-60427-123-2

Printed and bound in the U.S.A. Printed on acid-free paper.

10 9 8 7 6 5 4 3 2 1

Library of Congress Cataloging-in-Publication Data

Names: Swarts, Nigel, 1981—author. | Dixon, K. W. (Kingsley W.), author.
Title: Conservation methods for terrestrial orchids / Nigel Swarts and Kingsley W. Dixon.
Description: Plantation, FL : J. Ross Publishing, 2017. | Includes bibliographical references and index.
Identifiers: LCCN 2017014233 (print) | LCCN 2017016762 (ebook) | ISBN 9781604277791 (e-book) | ISBN 9781604271232 (hardcover : alk. paper)
Subjects: LCSH: Orchids. | Orchids—Conservation.
Classification: LCC SB409 (ebook) | LCC SB409 .S93 2017 (print) | DDC 635.9/3472—dc23
LC record available at https://lccn.loc.gov/2017014233

Direct all inquiries to J. Ross Publishing, Inc., 300 S. Pine Island Rd., Suite 305, Plantation, FL 33324.

Phone: (954) 727-9333
Fax: (561) 892-0700
Web: www.jrosspub.com

CONTENTS

FOREWORD

Orchids attract superlatives. Their family contains more species than any other plant family. At last count, there were around 25,000 of them and there are almost certainly more waiting to be identified—including many terrestrial orchids, the subject of this book. They are the most varied of plants. Some are so small they weigh only a few grams and produce flowers no bigger than match-heads. Others are giants—single specimens of which may weigh as much as five tons.

They are among the most widely distributed of plants. Some grow in bogs and others in near deserts. Many bloom in the high canopy of tropical rain forests; while others survive on bare rock. One genus has lost all its leaves and lives underground where it is pollinated by termites—the only plant yet discovered to be so. They are the very pinnacle of botanical evolution.

Such superlatives can be validated both statistically and scientifically, but orchids also appeal to the emotions. They are surely the most glamorous of plants. According to the dictionary, glamour is a magical, enchanting, and voluptuous quality. Exactly so—if any plants can be described as voluptuaries, it must surely be those orchids that produce flowers that mimic the appearance and smell of female insects with such extraordinary accuracy that male insects attempt to copulate with them and in doing so, pollinate them.

Orchids, in the opinion of many, produce the most spectacular and visually ravishing of all blooms. Until recent times, orchids could claim yet another superlative—that of being the most difficult of all plants to raise from seed. That characteristic, combined with the passion with which so many orchid enthusiasts wish to cultivate them, caused orchids to be the most expensive of all plants. In nineteenth-century Britain, single specimens changed hands for thousands of pounds. And that, in turn, has led orchids to acquire the saddest of superlatives. They are now among the world's most threatened plants. Ruthless collectors have stripped them from the forests of the world with such thoroughness that today many hundreds of species are in danger of extinction.

This authoritative and magnificent book is the work of many years of two well-respected scientists in the field, who clearly share a lifetime passion for terrestrial orchids. It describes breakthrough orchid research that now makes it possible to propagate even the most recalcitrant of species. This book includes case studies of conservation success and technology advances written by experts in their field from around the world. It deals with the practical problems of conservation and explains what actions must be taken if such work is to succeed. In doing all of these things, it dispels one of the mysteries that has surrounded orchids for so long. But that, as the following pages prove, not only enhances the claim of orchids to be among the most wonderful and astonishing of all plants, it also provides real hope that they still exist, in all of their diversity, complexity, and beauty, to delight generations of orchid lovers yet to come.

Sir David Attenborough
Richmond, United Kingdom

ACKNOWLEDGMENTS

The concept for this volume was drawn from an Orchid Conservation Techniques course hosted by Kings Park and Botanic Garden for the inaugural International Orchid Conservation Congress in Perth in 2001. We acknowledge the many contributors to the original training manual which has now been rewritten, updated, and expanded with the addition of sections on in situ conservation, reintroductions, pollination, and genetic testing approaches.

To the many colleagues who generously provided case studies showcasing the methods in practice around the globe, we are extremely grateful. We express our sincere gratitude to Tom Montgomery and Shiloh Isham for the preparation of the artwork contained in this volume, bringing to life the methods in a way that is both artistically outstanding and clear to the reader.

Additionally, we thank the many people who contributed images to accompany the text for both the case studies and chapters. Sources for all images are provided within the image captions. Images without source recognition were provided by the authors.

We are indebted to those who agreed to referee chapters herein: Hanne Rasmussen, Ryan Phillips, Suzanne Leis, Viv Muller, Phillip Seaton, and Joan Booth.

Finally, we thank Gwen Eyeington and Steve Buda from J. Ross Publishing for their editing and handling of this volume and bringing it to publication.

This volume is dedicated to an outstanding orchid enthusiast, conservationist, practitioner, and astonishing human being. Holger Perner died before seeing this volume published. The section by him and his wife on their remarkable work in Chinese orchid conservation is living testimony to the life of a man devoted to orchids. We are honored to dedicate this work to his memory and to support Wenqing, his wife and orchid partner, who continues Holger's nursery and orchid conservation work.

Nigel Swarts
Kingsley Dixon

ABOUT THE AUTHORS

Nigel Swarts, Ph.D.

After being introduced to the orchids of the Swan Coastal Plain in Perth, Western Australia by his father, Nigel developed a keen interest in the Orchidaceae. The rediscovery of a threatened spider orchid in bushland close to his childhood home paved the way for a future in orchid conservation research. Nigel received his undergraduate degree in Environmental Biology at Curtin University. He then completed his Ph.D. with distinction on the conservation of critically endangered orchids at the University of Western Australia. Nigel's research has led to a better understanding of the role of mycorrhizal specialization in the ecology and rarity of the Orchidaceae. His work has contributed to the development of new approaches to the conservation and recovery of terrestrial orchids based on key biological and ecological requirements for orchid survival.

During his first post doctoral position, Nigel embarked on an ambitious project to conserve all of Western Australia's terrestrial orchid seed and fungi in perpetuity. Nigel managed the orchid research programs of the Kings Park Botanic Gardens, supervised students, and trained teams of volunteers in the ex situ culture and propagation of Western Australian orchids using many of the techniques described in this book. In 2009, Nigel moved to Tasmania and established a similar orchid conservation and research program at the Royal Tasmanian Botanical Gardens, focusing on Tasmanian endemic species. He runs an ex situ conservation program where volunteers are trained in orchid conservation and propagation methods. He authored the multispecies recovery plan for all of Tasmania's threatened orchid species and continues his research into the mycorrhizal relationships of Australian terrestrial orchids on a part-time basis.

Nigel is currently a Research Fellow at the Tasmanian Institute of Agriculture at the University of Tasmania. In addition to his research, Nigel supervises postgraduate students in topics ranging from tree physiology to cider production and terrestrial orchid ecology. Nigel is also an active member of the Tasmanian Threatened Species Committee.

Kingsley Dixon

Kingsley Dixon is the Curtin University Professor in conservation science and restoration ecology, Perth, Western Australia, Research Associate at Missouri Botanic Garden, and Visiting Professor at The University of Western Australia. He is one of Australia's leading botanists who found his passion for orchids as a small child and would often collect them with his parents on bush walks. His professional and scientific interests in orchids developed while obtaining his Ph.D. at The University of Western Australia when he worked on the ecology of native southwest Australian orchids. Later, he expanded this into propagation, mycorrhizal, pollination, and conservation research when he established the world-acclaimed Kings Park Botanical Research Facility in the State Botanic Garden in Western Australia. Since then, Kings Park has produced more than sixty Ph.D.' s in botany and the biological sciences, and is home to an internationally recognized orchid research group that has made many major advances, including the development of novel tissue culture procedures for rare orchids and the first cryo-banking methodology for rare orchid seed. Kingsley was hired by Curtin University in 2015 and continues to lead major conservation and restoration programs that include orchid biology, ecology, and conservation research.

Kingsley has developed extensive research collaborations in conservation and restoration science in China, Southeast Asia, Europe, and the Americas. This has included programs to save rare and threatened slipper orchids in Indonesia and developing conservation propagation of rare European terrestrial orchids. Through these partnerships, he has developed a global network of orchid researchers and collaborators that includes significant input from citizen science programs from local to international orchid groups, including the Hardy Orchid Society of the United Kingdom.

With an incredible drive and enthusiasm for conservation and restoration of native ecosystems in Australia and around the world, Kingsley has published over 400 peer reviewed papers and 13 books; and has been recognized with several awards, including the Australian Orchid Foundation Award of Honour, the international Linnean Medal in Botany, and the Western Australian Scientist of the Year (2016/17). His passion for orchids and their conservation led him to establish the well-known International Orchid Conservation Congress series in 2001. Currently, Kingsley holds positions on international boards and commissions, including co-chair of the Orchid Specialist Group of the IUCN, Chair of the Research Committee of the Australian Orchid Foundation, Chair of the International Network for Seed-based Restoration, Chair of the Society for Ecological Restoration Australasia, and Board member of the Society for Ecological Restoration.

1

INTRODUCTION TO ORCHID CONSERVATION

From the Greek god Dionysus to Darwin, orchids have fascinated countless generations with their inescapable beauty and beguiling forms. Whether it is their scent, sexual allure (for both insects and humans), or astonishing ability to grow in some of the most remarkable habitats on earth, it is no doubt that the orchid family—more than any other family of plants—has captured the human imagination. Their sheer diversity, with more species than any plant family, means they might not be that common; but there is an orchid likely to grow in just about any environment or habitat—from sub-Antarctic islands to searing granite rocks in the outback of Western Australia.

With so much diversity and exuberant evolutionary success, why are orchids also the single most endangered family of plants on earth? The answers to this are to be found in the pages of this book, as well as practical approaches currently available in orchid conservation techniques and technology. The concept for this book came from the inaugural International Orchid Conservation Congress (IOCC) held in Perth in 2001. Over time, and four IOCC conferences later, the book has grown to cover all aspects of orchid conservation with a focus on making orchid conservation technology accessible to all.

So why do we need a book on orchid conservation techniques? With 12% of the world's plants facing imminent extinction, and another 40% predicted to be extinct by 2050, never before has global biodiversity been under such threat. As we face the prospect of the sixth great extinction event, the challenge remains for conservation biologists, land managers, orchid enthusiasts, and growers to coordinate and link science and practice for more effective and targeted conservation outcomes. The task is large, and the time is limited. Orchid growers, admirers, and scientists have a special role to build the conservation technology toolkit and use these skills to conserve and reestablish wild orchid populations. Importantly, when engaging with stakeholders—whether they be landholders with orchids on their land, decision makers, or simply community members—dialogue is needed to forge links to deliver a level of conservation that will stem the impending tide of global extinctions threatening to engulf this remarkable family.

1.1 WHY ARE ORCHIDS RARE?

Orchids have significant below- and above-ground interactions with the living and abiotic environments. Few other families have such a diversity of synergies with other organisms: below ground, fungi and bacteria inhabit roots, rhizomes, and stems to assist with nutrient acquisition, immunity, and germination; above ground (and below ground in Australian underground orchids), pollinators from birds to bees, wasps, moths, and butterflies are sometimes rewarded; more often than not, they are duped into shifting orchid pollen around the landscape. With over 25,000 species globally, these critical biotic associations may have played major roles in driving the astonishing diversification of the family, but at the same time, the dependence of orchids on these interactions for sexual reproduction and seed germination through to maturity often results in populations that are highly dispersed, limited in abundance, or vulnerable to even the slightest change in habitat conditions. As a result, such species face heightened extinction risk, particularly from natural climatic shifts and human-induced change.

1.2 WHAT THIS BOOK MEANS TO ORCHID CONSERVATION

This book is structured to provide the reader with the capacity to develop a conservation program primarily for terrestrial orchids; nevertheless, many of the techniques apply equally to epiphytes. It would take an encyclopedia of books to review orchid science and conservation practice; however, this book is a navigation aid to getting started. Though each of these chapters provides stand-alone guidance, collectively they enable practitioners to undertake what were previously considered complex scientific procedures. These chapters are enriched with invited *Case Studies* that are relevant to illustrating key principles, along with success stories in orchid conservation that were written by world leaders in orchid conservation practice and science.

1.2.1 Chapter Outlines

Monitoring is covered extensively in Chapter 2. The how, when, and what to monitor has been integrated into demographic studies—understanding the recruitment cycle from seed germination to adulthood to senescence of the orchid plant. These life-stage transitions provide a firm foundation for understanding recruitment bottlenecks and where to prioritize conservation actions. Incorporated in Chapter 2 are four salient case studies that were provided by experts in the field. They show the conservation benefits of monitoring and demography studies including analytical approaches as applied by Raymond Tremblay and his colleagues. Chapter 2 concludes by demonstrating another important use of monitoring and demographic data in developing protective policies through processes such as *Red Listing* under the International Union for Conservation of Nature (IUCN).

Chapter 3 covers core aspects of orchid mycorrhizal ecology relative to conservation actions and planning. Though mycorrhizal ecology of orchids is one of the most wide-ranging and highly published aspects of orchid biology (outside of the orchid taxonomic literature), practitioner benefits of these research studies is often limited. In this chapter, we synthesize approaches used for mycorrhizal isolation and culture using appropriate media. A special case study on the newly emerging, but highly specialized field of isotopic discrimination is presented by Gerhard Gebauer and Julienne Schiebold. This tool enables understanding of orchid nutrition, with the case study providing a readable and up-to-date account of how to use this potent tool to understand the degree of mycorrhizal dependency in orchids—with some surprising outcomes.

Ex situ conservation methods are explored in detail in Chapter 4, where orchid seed sowing techniques are outlined—ranging from sterile to semi-sterile germination techniques and nursery sowing. Though techniques in orchid seed sowing harken back to the pioneering work of Lewis Knudson in the 1940s, techniques and media to solve issues with seed germination continue to develop and improve. This is illustrated by the outstanding case study of the industrial scale propagation of *Cypripedium*, in both China and Europe, which was once considered to have seed that is deeply dormant and difficult to germinate. Even though the mycorrhizal fungus remains elusive to culture, the techniques developed by commercial and conservation practitioners (often involving green pod culture or, for mature seed, exposure to oxidizing agents to remove testa-imposed dormancy) are now reliably producing many seedlings.

Following on from Chapter 4 is an exploration of *in situ* sowing and soil baiting techniques in Chapter 5. Based on the pioneering work of Hanne Rasmussen and colleagues, the use of seed baits placed into soil has revolutionized our capacity to determine mycorrhizal distribution in soil as well as isolating ecologically competent fungi from developing protocorms. Soil baiting is also a tool for auditing soils to assess orchid recruitment hotspots with the technique modified to also assess recruitment niches for epiphytic species.

Chapter 6 is an overview of the techniques for orchid culture, focusing on symbiotic approaches to the culture of terrestrial taxa where conservation issues are the greatest. Epiphytic orchids, on the other hand, are readily and reliably propagated—by the millions! Just walk into any big box store, florist, or nursery, and the abundance and diversity of epiphytic species on sale is a clear sign that culture techniques are well resolved for this group of orchids. Many terrestrial taxa are routinely propagated and grown with nurseries specializing in their local terrestrial species. A fascinating local capacity building case study showcases the use of these techniques in an integrated orchid conservation propagation project in Africa presented by Vincent Droissart and Tariq Stévart. However, some species remain stubbornly elusive to long-term culture, such as the sexually deceiving genus *Drakaea*, where it's distinct and unusual mycobionts decline under pot culture conditions.

A key tool in modern day orchid conservation is the translocation of propagated plants to safe or restored sites. In Chapter 7, successful translocation approaches are described with a short review on the principles of restoration ecology. Five case studies present success stories in orchid translocation including involving undergraduate students in North America by Lawrence Zettler; the role of the Atlanta Botanical Gardens in restoring habitats for orchid reintroduction; a Ph.D. research project in southeast Australia; restoration of *Cypripedium* in the mountains of Beijing; and finally, how an orchid enthusiast managed to repopulate meadows in Scotland with common and rare local orchids. Strategies, decision support tools, and policy frameworks for translocations are also presented in this chapter and in other parts of the book.

Orchids share a unique and unusual relationship with their pollinators with a high level of specialization by the orchid on particular pollinators explaining, in part, the extraordinary diversification of the family. Chapter 8 provides an overview of orchid pollination syndromes drawing upon the remarkably diverse literature on the subject—from apomictic taxa through to orchids engaging in sexual deception. We show how to assess pollinator presence and efficacy through floral baiting and how to interpret these findings in conservation planning when translocations occur into new habitats or restored locations. A case study by Xiaokai Ma and Yi-bo Luo demonstrate the role of flower color signal mimicry in pollinator attraction and pollination success in the slipper orchid *Paphiopedilum micranthum*. Another by Colin Bower demonstrates the implementation of targeted field research approaches to the study of sexual deception in orchids.

Chapter 9 outlines genetic approaches for all aspects of orchid biology and conservation—drawing upon the wealth of molecular approaches to assess genetic diversity in orchid populations to describing fungal diversity. Detailed approaches and methodologies are provided which, though of a highly technical nature, will be of great value to research students and laboratories with molecular capacity. Included in this chapter are case studies of how molecular techniques can facilitate taxonomic resolution in difficult genera, assist with species identification of orchid tubers that are being illegally traded, and finally, target important genotypes for the cultivation of the enigmatic *Cypripedium calceolus*.

A fundamental tool in conservation where rare and threatened species are involved is the use of *ex situ* conservation through conservation germplasm storage (mycorrhiza, seed, and living collections). Though many orchid seed banks exist throughout the world, Chapter 10 highlights that standard approaches, though generally applicable to many species if not carefully assessed, can lead to seed viability decline in some species. The chapter provides an overview of approaches including how to secure high quality collections of orchid seeds, as well as innovations in alginate encapsulation of seed and mycorrhizal fungus. Storage of mycorrhizal fungi and maintenance of living collections are also included in this chapter. Case studies for a number of *ex situ* orchid conservation programs are presented by Jennifer Cruse-Sanders and Dennis Whigham.

1.3 EPILOGUE ON THE FUTURE OF ORCHID CONSERVATION

The need for orchid conservation is paramount if we are to leave to future generations the rich and wildly fascinating orchid legacy we all enjoy today. Without the effective conservation actions outlined in this book, threatening processes will continue to militate against the survival of rare orchids, resulting in their continued degradation and inevitable extinction.

Integrated conservation of orchids (where often disparate scientific and practical approaches are melded for the purpose of a conservation outcome) provides a balance between the priorities of the need for scientific research and the need for urgent action in light of the large and growing number of rare and threatened orchid species (see Figure 1.1). However, despite the clear need for such an approach, there is relatively little work published on integrated conservation of rare plants, let alone orchids, that encompass genetic analysis, biological and ecological research, and experimental reestablishment of populations. While these data are substantially less in the context of terrestrial orchids, the importance of such knowledge is clear and the time for action is *now*.

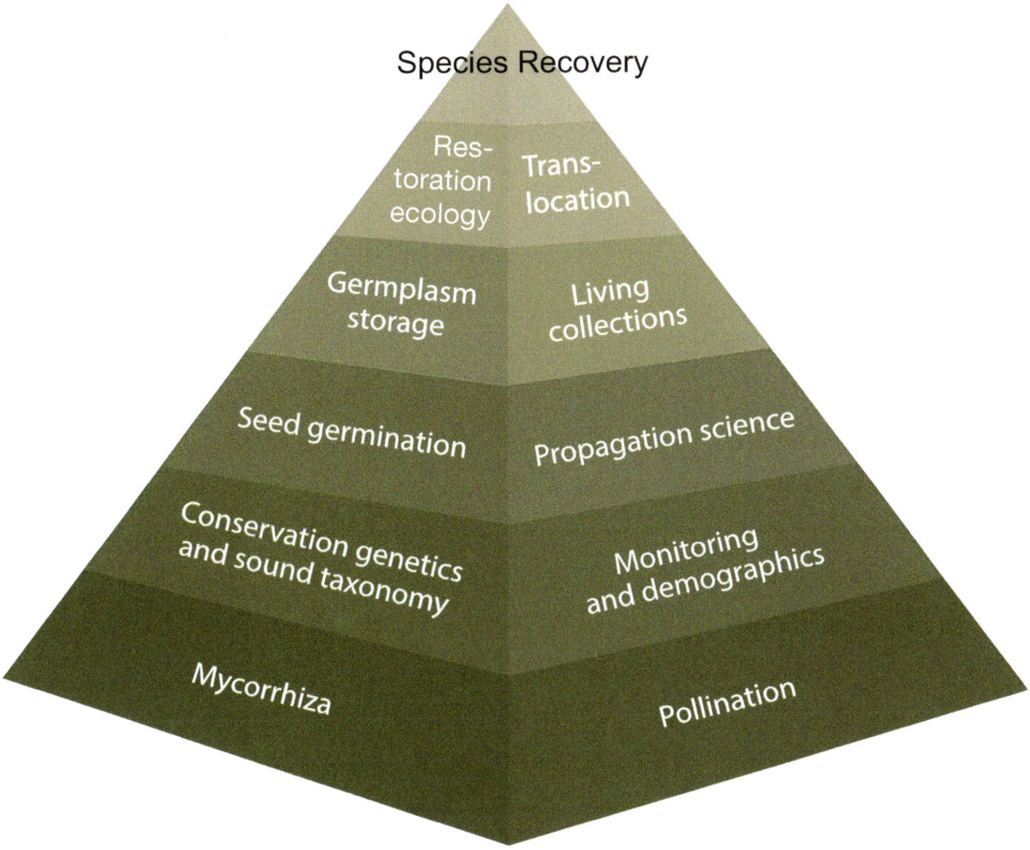

Figure 1.1 Concept model of integrated conservation relevant to orchids where individual scientific disciplines are linked to deliver effective and practical conservation outcomes

2

IN SITU MONITORING, DEMOGRAPHICS, AND CONSERVATION INSTRUMENTS

2.1 BACKGROUND

In this chapter we summarize the approaches for monitoring and demographic investigations relevant to orchid conservation. The chapter also provides an overview of conservation instruments including the important processes of the International Union for Conservation of Nature (IUCN) Red Listing (that depends in large part on monitoring and demographic information), complimented by recovery planning relevant to developing effective orchid conservation plans and programs.

Monitoring threatened orchid populations is the systematic collection of data that will detect changes in plant abundance over time and ultimately improve conservation management. Monitoring is a method for identifying change in recruitment and death in a population that determines if a population is declining, static, or increasing in size. How these factors are measured varies along a monitoring continuum. Monitoring data comes in all shapes and sizes from the simplest (presence-absence record of a species at a site over time) to labor-intensive demographic monitoring that collates detailed life history information about individual plants. The type and intensity of monitoring depends upon the management questions being asked and may be species or population specific.

Demography (the study of the statistics of births, deaths, disease, etc.) is the primary way ecologists analyze changes through time in the condition of a population (Hutchings 1990). Annual counts do not provide sufficient information on which to base management decisions or to assess the success or failure of management actions (Hutchings 1990). Therefore, demographic studies involving multiple annual censuses are required to determine ecological parameters including life span; rates of mortality and reproduction; age of reproductive maturity; and response to grazing, fire, or other management activities.

2.2 MONITORING METHODS

Terrestrial orchid demographic studies can be difficult because in many species the re-emergence of plants above ground in successive seasons is not guaranteed. The propensity for orchids to produce a leaf and flower is dependent on a range of factors (most not well understood)—including density of competing vegetation and climatic conditions. For epiphytic and lithophytic species, though they are evident year round, the plants are often in inaccessible parts of the tree canopy or rocky cliff faces. Consequently, it can be difficult to trace the progress of many orchids through their life cycle.

The aim of population demographic monitoring is to follow individual plants from one year to the next so that a record of their reproductive effort and life history traits (e.g., dormancy patterns) can be recorded (baseline data collection) (see Case Study 2.1). Population monitoring can be achieved through baseline surveys using the methods described in the following text or through extension surveys/searches in a priority habitat (e.g., a previously unsurveyed habitat suitable for orchid presence, a habitat experiencing a recent disturbance event, and opportunistic surveys in new locations). Baseline surveys are necessary to accurately map the location of subpopulations and describe habitats and threats. Improving baseline data enables a better understanding of the current extent of species' distribution and allows a more accurate determination of a potential habitat for extension searches. A template for recording information following a survey is provided in Table 2.1.

Baseline data can be used to identify critical management actions for existing populations or new populations identified through extension surveys. Landowners, environmental consultants, conservation managers, and orchid enthusiasts (citizen scientists) are encouraged to record and report on orchid subpopulations. When new subpopulations or threats are detected, the management procedure detailed in Figure 2.1 can be followed.

When conducting demographic studies of orchids, it is essential that the precise locations of plants be recorded. The following methods originally described in TSS (2006) are some examples of how to monitor orchid populations. Each method will require tailoring to suit a particular site and population characteristics; however they provide a guide for designing monitoring systems for any orchid population.

2.2.1 Permanent Pegs

One of the simplest methods in demographic studies is marking the location of the orchid with a permanent marker (see Figure 2.2). A metal peg is recommended of a suitable size to assist with relocation placed in the soil approximately 10 cm away from the plant. This is to avoid damaging the underground organs of the plant. Stainless steel should be used to avoid potential rust issues. It is essential that the direction the peg is positioned from the plant is systematic and recorded so that plants in close proximity do not get confused. A permanent tag should be fixed to the peg as an identifier and can contain information such as the year it was first recorded along with markings for when the plant has appeared in subsequent seasons. A metal detector can be used to locate the steel pegs if the layer of leaf litter or undergrowth thickens or increases over time.

Table 2.1 Example of a record template for recording demographic information about an orchid population

COLLECTION RECORD		
COLLECTOR INFORMATION		
Name:	E-mail:	
Date:	Phone:	
SPECIES INFORMATION		
Genus:	Species:	
Common name:		
No. of individuals:	Area of occupancy (m^2):	
Habit (flowering, vegetative, fruit, bud): Relative proportions of each:		
Has a specimen been collected?	Voucher no:	
LOCATION		
Location:		
GPS coordinates:		Datum:
Aspect:	Slope:	Elevation:
Tenure:		
SITE DESCRIPTION		
Description of vegetative community:		
Associated species:		
Geology (soil texture, rock cover):		
Landforms:	Drainage:	
OTHER NOTES		
(e.g., fire history, land use, threats, disturbance, flower color, presence of fruit/flowers, evidence of recruitment, population demographics, etc.)		

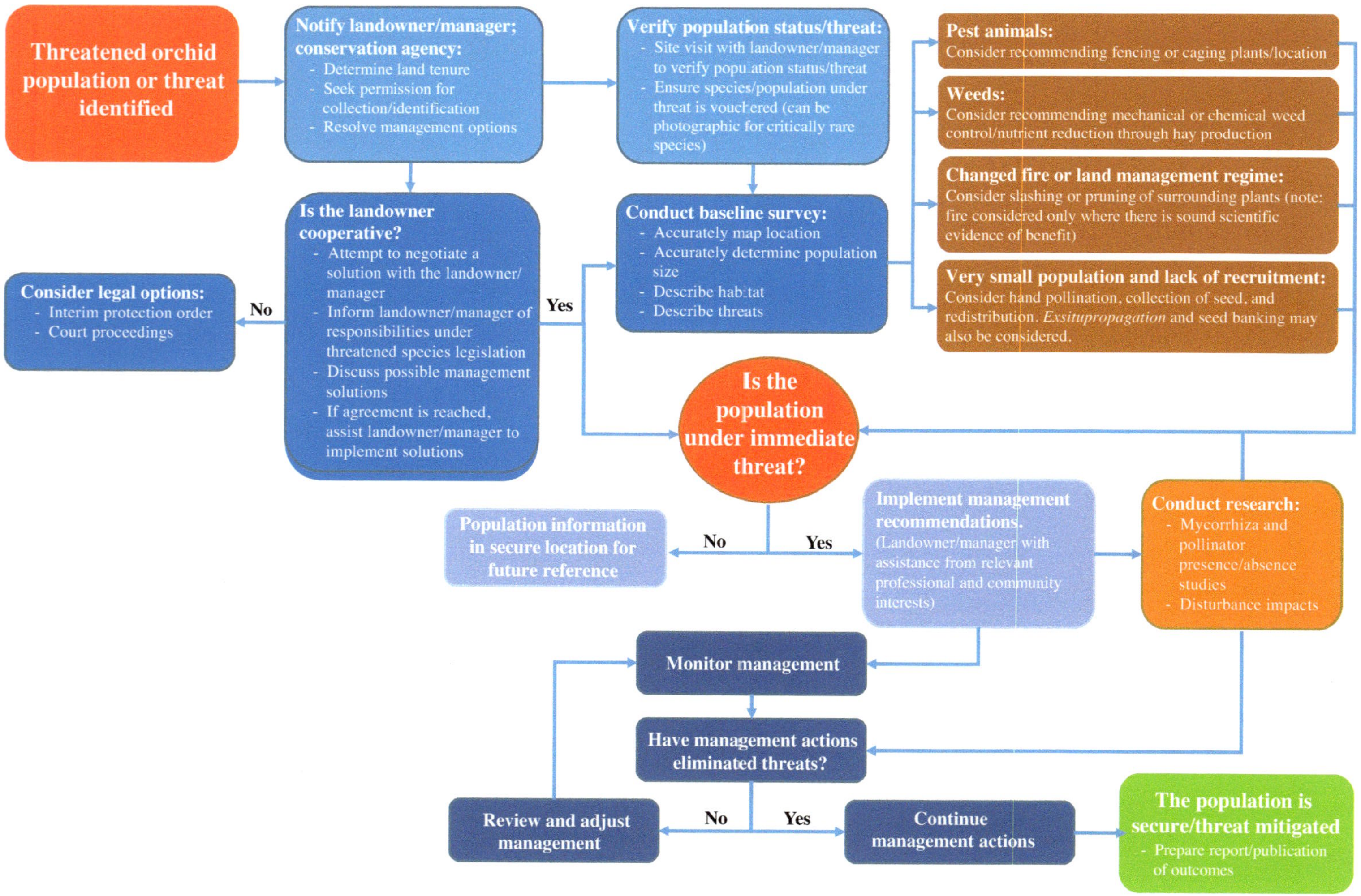

Figure 2.1 Conceptual framework for orchid conservation actions with key activities/concepts and decision points leading to effective site-based and long-term conservation outcomes

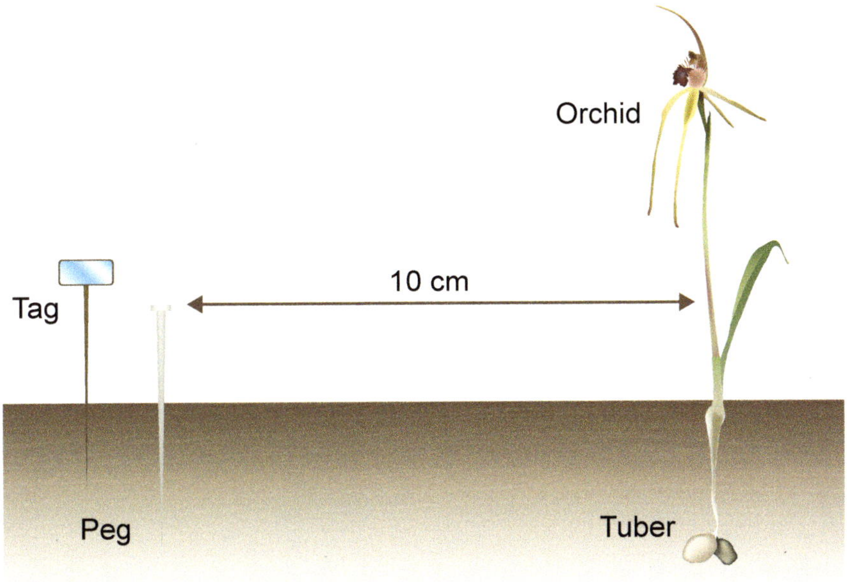

Figure 2.2 Permanent monitoring tag

2.2.2 Transect Method

A transect line dissecting a population is delineated using two permanent pegs (see Figure 2.3) and measuring tape between them. The locations of plants are mapped by measuring both their distance along the transect line and their distance out from the transect line. Set a limit to the distance from the transect so the plot has a defined area. Plant positions can also be marked with a peg so their approximate location can be identified even if they fail to produce a leaf or flower. This method is particularly effective for populations that are widely dispersed.

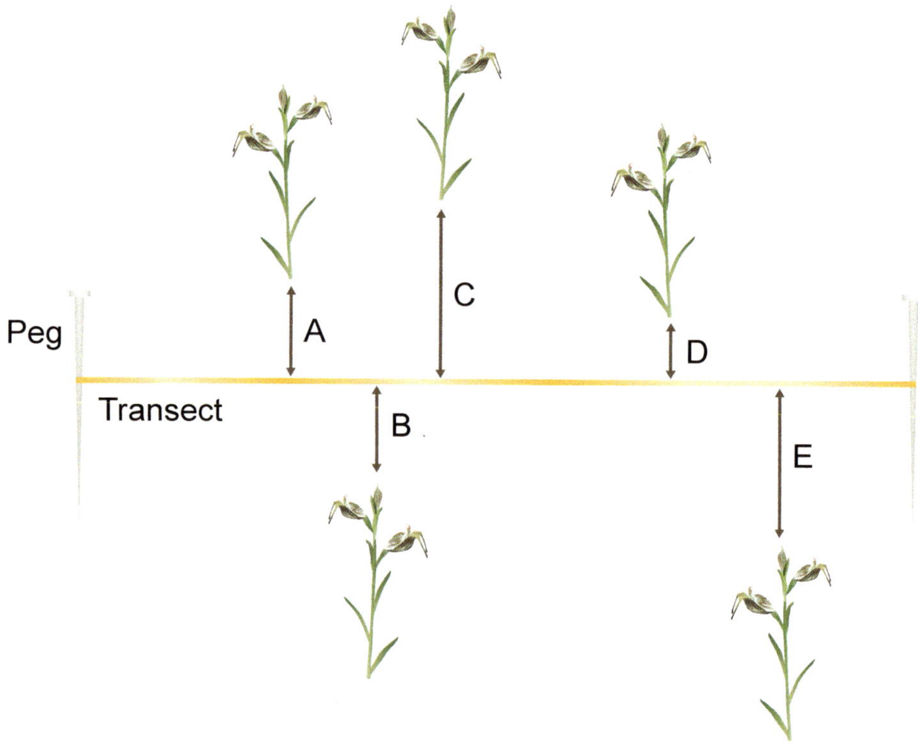

Figure 2.3 Transect method

2.2.3 Plot Method

Square plots of a defined area (e.g., 5 m × 5 m) can be marked with permanent pegs (see Figure 2.4).

Plant locations can be mapped by measuring its distance from two of the permanent pegs (i.e., Peg A and Peg B). The formula below is used to plot the location of the plants on the X and Y axis relative to the position of the pegs. This method is effective for a population that occurs in a single or in multiple small, well-defined patches. Formulae are described below:

X coordinate = (D1)*COS(90*(π/180)-((ACOS(((Distance between pegs2) + D12-D22)/((Distance between pegs)* D1)))))

Y coordinate = D1 *SIN(90*(π/180)-((ACOS (((Distance between pegs2) + D12-D22)/((Distance between pegs2)* D1)))))

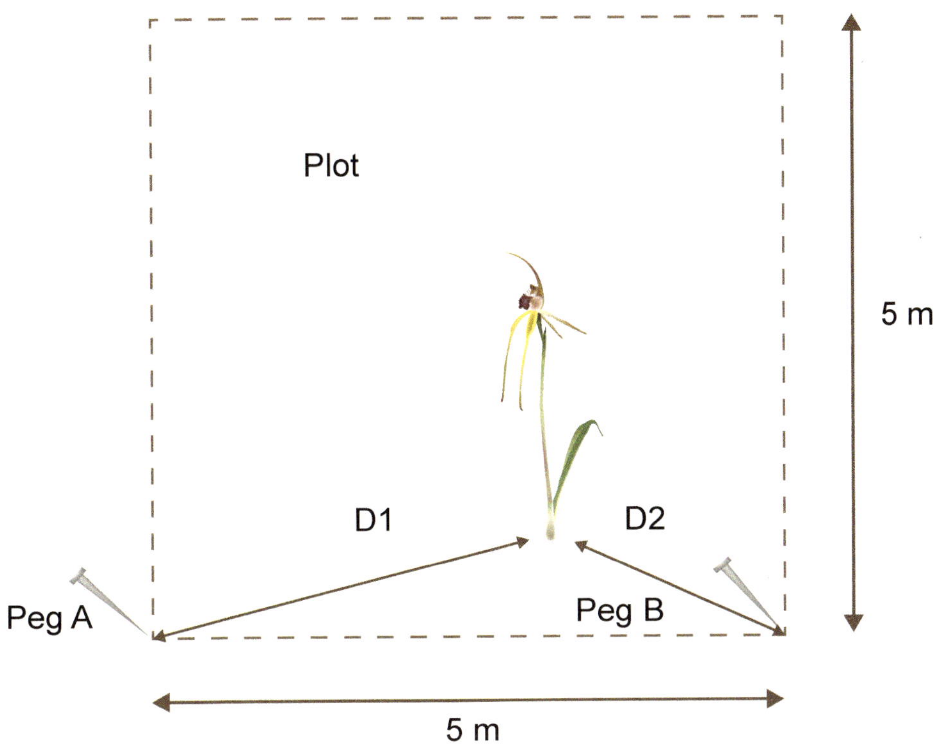

Figure 2.4 Plot method

2.2.4 Mapping and Recording Orchid Recruitment

Naturally recruited orchid seedlings often have a high mortality rate. Seedlings germinate and produce leaves but many die before they reach reproductive maturity. Management practices like trimming competing vegetation, careful and light scarification of soil, and distribution of seed around parent plants (discussed in Chapter 4) can increase the likelihood of seedling establishment. Recording the location of new seedlings is complicated by the difficulty in identifying species from just their vegetative structures that for many genera are indistinguishable between species at the seedling stage. Marking each seedling with a peg may not be possible without potentially damaging other plants. It is advisable to record the number of seedlings, but to avoid mapping or marking their location with pegs until they flower or have diagnostic vegetative states.

The development of seedlings can be monitored by marking their location on a clear plastic sheet located with permanent pegs defining a 1 m × 1 m area centered around the parent plant (see Figure 2.5). Seedlings' positions are marked on the sheet by standing above the seedling area and noting their position on the transparent sheet. The information is then digitally stored for future reference. A digital photograph of the plastic sheet should be taken, and using rectifying software the number and location of seedlings can be effectively recorded.

Figure 2.5 Monitoring seedlings

Case Study 2.1

Extraordinary variation in vegetative and floral characteristics among years in a terrestrial herb

by Raymond L. Tremblay[1] and Andrew Bould[2]

In situ conservation requires only one basic aim, the assurance of long-term survival of the targeted species. Although the concept is simple, predicting the likelihood of long-term survival is far from easy (Caswell 2001; Coates et al. 2006; Schodelbauerova et al. 2010; Tremblay et al. 2009a; Tremblay et al. 2009b). Long-term monitoring is required through a process of mark-*recapture*, where individuals are identified, marked, and resurveyed across a time frame that is concordant with the life span of the species (Caswell 2001). In addition, when information on the vegetative and morphological characters are collected, this biological information can be invaluable and may potentially suggest other venues of study or alternative management strategies.

Some of the advantages of a long-term survey resides in the ability to evaluate changes in time which is likely a consequence of environmental change, but can also be a consequence of an evolutionary process such as natural selection or genetic drift (Tremblay et al. 2005) or, more likely, an interaction between both of these. To illustrate the value of long-term surveys, we provide an example with the Australian herbaceous perennial terrestrial orchid, *Caladenia valida* (see Figure 2.6). The data was collected by Andrew Bould from plants on Raymond

Figure 2.6 *Caladenia valida* developmental stages: (a) vegetative leaf; (b) flower bud; (c) flowering plant; and (d) ripe seed capsule (photos: Andrew Bould).

Island, Victoria and shows the value of how citizen science can actively assist with meeting the goals of long-term surveys. Moreover, the information that was gathered demonstrates the importance of long-term continuous sampling of the same populations and individuals. Bould tracked over 300 unique plants throughout the period of 2004 to 2015 measuring emergence, leaf and flower size, fruit set, and herbivory.

As a consequence of this long-term survey, the study found that plant size varies greatly among years. Examples of this include changes in mean size and dispersion (variance) of vegetative (see Figure 2.7) and floral (see Figure 2.8) characteristics among years. In Figure 2.7, the mean overall height is shown to differ among years, while in Figure 2.8, the length of the lateral sepal also varies substantially among years. What is the cause of this variation in time?

From a simple survey we cannot attribute the cause of the variation, however, we can speculate, and the cause of this variation can be tested using appropriate *in situ* data collecting and experimental approaches. The amount of variation in this species is outstanding and long-term studies of other species or orchids should be considered. Will we find similar patterns

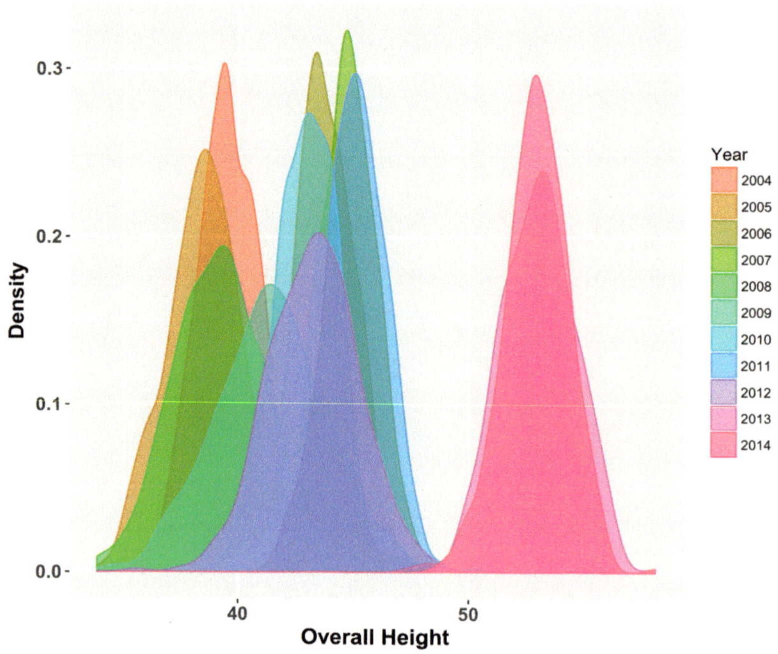

Figure 2.7 Bayesian posterior probability distribution of overall height (cm) of *Caladenia valida* from 2004 to 2014. The overall height of plants in the last 2 years were much larger than previous years. Note that the mean and the shape of the distribution differed dramatically among other years also.

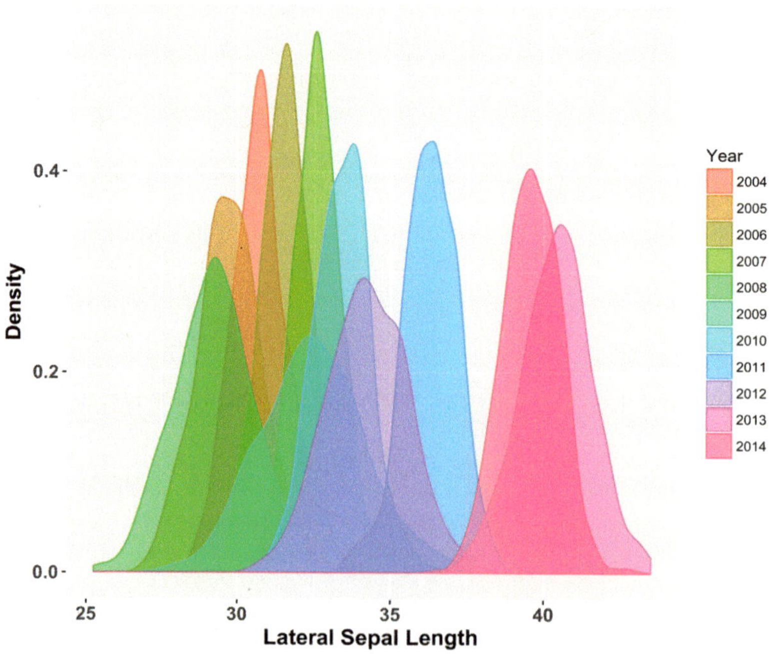

Figure 2.8 Bayesian posterior probability distribution of the length of the lateral sepal (mm) of *Caladenia valida* from 2004 to 2014. Notice the lack of overlap among many of the years.

among other orchid species, and can we assign some extrinsic environmental or evolutionary variables that can start explaining this large variation? Only after further long-term studies on other orchids will we begin building the knowledge base for providing more broadly applicable principles.

[1]University of Puerto Rico, Center for Applied Tropical Ecology and Conservation, San Juan, Puerto Rico, USA

[2]109 Western Boulevard, Raymond Island, Victoria, Australia

2.3 ESSENTIAL PARAMETERS IN DEMOGRAPHIC STUDIES

Essential parameters to be recorded for each plant during a demographic study are listed below:

- Vegetative or flowering status of the plant and the stage of floral development
- Number of flowers on each plant
- Number of flowers pollinated
- Grazing pressure, attacked by insects or disease (specifying the affected parts)
- Number of seedlings around the parent plant
- Microhabitat of the plant (e.g., whether it is growing out of moss, in the shelter of other plants, in bare soil, depth of leaf litter, etc.)
- Notes regarding the specifics of the site and species (e.g., holomycotrophic orchid and co-habiting tree species)

Some other parameters that can be measured include:

- Scape length
- Leaf length
- Soil disturbance, if evident, including previous fire history
- Seed-pod development—repeated monitoring throughout the flowering period to determine the proportion of flowers successfully pollinated and the proportion of pollinated flowers producing seed

Case Study 2.2

Modeling approaches for conserving a rare temperate terrestrial orchid

Patti Vitt
Chicago Botanic Garden, Glencoe, Illinois, USA

The woodland terrestrial orchid, *Isotria medeoloides* (Pursh) Raf. is a federally threatened species in the U.S. that is found in small, isolated populations in eastern North America (see Figure 2.9). The species is a herbaceous geophyte, arising from perennating buds on the growing edge of a rhizome, which lie approximately 2–5 cm below the soil surface. The single (rarely, two) flowers are small, green, apparently without scent or nectar (Mehrhoff 1983), and are self-pollinating (Vitt and Campbell 1997). Vegetative reproduction is rare to infrequent (Mehrhoff 1983), resulting in populations of widely scattered individuals. Individual plants may fail to produce an aboveground shoot in some growing seasons (Vitt 1991).

Figure 2.9 (a) *Isotria medeoloides* in its natural habitat (photo: Aat Barendregt) and (b) Close-up of an *Isotria medeoloides* flower (photo: Hal Horowitz).

Permanent plots were established in 1986 at four sites in the southern regions of Maine in the U.S. Data on aboveground sprouting were taken twice annually in June (Census 1) and August (Census 2). One to three plots were positioned to capture the greatest density of individual plants at each study site, and plot size varied from 10 m × 10 m to 20 m × 20 m. Plots across all sites incorporated 255 individual plants when the study commenced. Each plant that produced an aboveground shoot was tagged and mapped. A plant was considered to be an individual (genet) if the base of the stem was greater than 5 cm away from the nearest neighbor; stems within 5 cm of each other were classified as a single plant (a genet with multiple ramets). Plants were categorized as emergent or not and as flowering or not (Census 1) and as emergent or not, and fruiting or not (Census 2). The diameter of the leaf whorl was taken during Census 2 as a measure of plant size. Levels of herbivory were recorded as none; light, if less than 50% of the leaf area had been removed; or heavy, if more than 50% of the leaf area had been removed. Data across both census dates were pooled and annual stage categories assigned (Dormant/Dead; Vegetative; Flowering Only; Fruiting).

In 1989, three of the four study sites experienced a sharp decline in emergent individuals, likely as a result of low winter temperatures concomitant with low snow pack. Plant size in each year was significantly correlated with annual stage (Vitt 1991). To explore the dynamic of extreme winter warming, an Integral Projection Model was created using data on 403 individual plants across all four sites. This model projected population growth rates significantly lower for every year during the study with low snow pack, leading to the conclusion that bare ground conditions in January increased mortality, depressing population growth. This study demonstrates the value of focused monitoring for managing and predicting conservation outcomes of rare orchids.

Case Study 2.3

Demographic monitoring of Prasophyllum limnetes *in North West Tasmania demonstrates how exogenous factors influence the presence and absence of orchids in a population*

Phil Collier and Robin Garnett
Rubicon, Port Sorell, Tasmania

Prasophyllum limnetes is currently known only from Rubicon in northwest Tasmania (see Figure 2.10a and b).

We identified a population in 2009 following a fire and commenced mark-*recapture* monitoring on all emerged plants, which continues to date. Mark-recapture monitoring involves

Figure 2.10 (a) *Prasophyllum limnetes*; (b) close up of *Prasophyllum limnetes* flowers; and (c) *Prasophyllum limnetes* leaf emergence, attempted flowering and fruiting by number of years since disturbance. The chart includes data for plants that were captured in the period 2009 to 2014, and not subsequently disturbed. Most plants are captured in flower, so this illustrates the decline in flowering post disturbance, which is typical of a disturbance-dependent species (photo: Phil Collier).

marking individual plants and visiting them three times a year to record systematic details about leaf emergence, flowering, and fruiting. The aim is to better understand population demographics, including a measure of population growth. However, the data collected has multiple uses. Initially, we used a morphometric analysis to confirm that the plants correspond to the species description, and that they differ from the similar-looking *Prasophyllum rostratum*.

Our analysis of monitoring data enables the profiling of leaf emergence and attempted flowering by the number of years since the disturbance. We now know that *Prasophyllum limnetes* is disturbance dependent, with flowering rates declining rapidly after the initial disturbance (fire). Furthermore, a few flower stems in later years shriveled prior to any flowers opening. We speculate this relates to crowding by surrounding vegetation that thickens rapidly following a disturbance and/or the prevailing dry conditions for this December-flowering species.

Using active management, we experimented with a 2012 slash for part of the area occupied by *Prasophyllum limnetes*. This small clearing created a magnet for the local herbivores, as our bracken markers were regularly broken. Responding to this, we instituted a caged-uncaged experiment in an effort to ensure that herbivores were not grazing on all of our precious plants. While the flowering in the slashed area was limited, some plants took advantage, and the shriveling of flower spikes was reduced.

In 2015, we again burnt the site of the *Prasophyllum limnetes* population, hoping to spread the resulting grazing pressure over a much larger area. Unfortunately, 2015 proved to be a drought year, with little flowering. Our six seasons of capture-recapture monitoring for multiple species has shown that the best flowering response is typically in the year after a disturbance (see Figure 2.10c).

How this data can be used for other disturbance opportunist species will require ongoing monitoring and further experimental work.

2.4 PHOTO POINTS AS TOOLS IN MONITORING

Anecdotal evidence suggests that the emergence of plants and flowering events can be influenced by the density and/or successional stage of the vegetation community in which the population exists. These factors will have an influence on the management of the population and its habitat. This highlights the importance of recording habitat conditions when conducting a demographic census. A photograph enables the quick recording of the density, structure, and if applicable, the successional stage of the vegetation community at the time of the census. A qualitative analysis can then be applied using the photograph to determine how the population changes with changes in habitat. A permanent photo point should be established at each site that is clearly marked for return visitation. Adjustments may need to be made in order to record the detail required—such as biomass accumulation or vegetation cover values.

Case Study 2.4

Ethical photography in a terrestrial orchid conservation context

Magali Wright and Nepelle Crane
Natural Resource Management South, Hobart, Tasmania

Over time, there has been a shift from collecting orchid specimens to collecting photographs, and while photography is significantly less damaging, there is still an impact—especially for rare species. There is anecdotal evidence that this impact has increased with the use of social media, where photographs and the location of flowering plants can be shared, leading to increased visitation to sensitive sites. Considering the irregular recruitment of many terrestrial orchids (Kindlmann and Balounova 1999; Pfeifer et al. 2006), a bumper flowering event can represent rare opportunities for recruitment, but also significant attention by enthusiasts keen on a photographic record.

Taking photographs of orchids is a powerful communication tool and motivator—it can ignite the first spark of interest in conservation. We all have a role in building a culture of ethical photography of orchids and other sensitive species. When undertaking a photographic expedition particularly involving a rare species, consider the following guidelines:

- Obtain land manager permission each time you visit—unauthorized access can impact the relationships that underpin conservation efforts.
- Avoid spreading pests, weeds, and diseases—before entering a sensitive site, make sure your footwear and equipment are clean of soil, water, plants, and animal material.
- Watch your step—damage (defoliation) of orchids can reduce plant vigor and reproduction success (Whigham 1990). For sensitive sites, consider if your visit is absolutely necessary and if so, watch where you put your feet, walking on exposed rocks where possible.
- Walk softly and stay on track—soil compaction can alter soil biology (Light and MacConaill 2007), and as orchids rely on colonization by mycorrhizal fungi for germination, this could impact recruitment. Disturbing the soil, grass, or cryptogram cover, can expose soil to sun, rain, and erosion—walk softly, stay on formed tracks, and avoid wet conditions (Light and MacConaill 2008).
- Avoid standing, kneeling, or lying directly around plants. If there is a clear conservation benefit to search off-track, keep to the same path when moving in and out of a site.
- Take care of surrounding vegetation—damage to surrounding vegetation can expose orchids to grazing, as there is evidence that orchids in open environments are more likely to be browsed (Garnett and Collier 2012). This may also impact pollinator interactions by reducing the viability of surrounding food plants—avoid removing or damaging neighboring plants to get the perfect photo.

The authors are working toward practical recommendations for ethical photography for Tasmanian native orchids and other location-dependent species; and we encourage others to work on similar initiatives in their local area. Let's help orchid lovers minimize their impact, while still enjoying these exceptional species.

2.5 RECOVERY PLANS FOR TERRESTRIAL ORCHIDS

In many countries rare and threatened orchids are protected by legislation that also identifies the need for a species recovery plan (RP) that addresses threatening processes affecting the ongoing survival of the species, so as to encourage natural regeneration to occur. RPs for threatened orchids are prepared within the framework designated by the overarching state or national government conservation authority (e.g., the Australian Commonwealth *Environment Protection and Biodiversity Conservation Act 1999*).

The objective of an RP is, during the term of the plan (often RP's are time-bound), to maintain or improve the conservation status of the species by ensuring the continued survival of the known populations, mitigating identified threats to populations, and particularly for orchids, supporting increases in the areas of occupancy and numbers of mature plants through translocation, once successful techniques are developed. RPs contain much of the information required to make informed decisions about the need for translocation—including species distribution; number, size and status of populations; previous research on the species; and complimentary information relevant to developing affective conservation strategies. Text Box 2.1 details the basic information contained within a threatened species RP.

Text Box 2.1

Recovery plans for threatened species

1. Background
   a. Nomenclature history, general introduction, distribution, and status
   b. Description of recent developments
   c. Detailed botanical description, flowering morphology
   d. Distribution and habitat
   e. Biology and ecology
   f. Research details with references, if any have been completed
   g. Description of identified threats to populations
   h. Summary table for each population including number of plants, population survey date(s), identified threats, location information
   i. Description of habitat necessary for the survival of the species and the species context within ecological communities
   j. Obligations and considerations: International; indigenous consultation; social and economic impacts; stakeholder and landowner consultative processes; recovery plan evaluation process (annual, five-yearly, or, *as required*)
2. Recovery actions
   a. Statement of specific objective and criteria for success and failure
   b. Identify recovery actions already completed or occurring in each population
   c. Identify future recovery actions required based on previous research, population, and habitat surveys, monitoring
   d. Budget for specific items including
      i. Habitat management
      ii. Consultation with stakeholders, including potential impacting activities

iii. Population monitoring and habitat surveys, including new locations
iv. Management of access to populations undertake weed, grazing control
v. Development of a fire/grazing management strategy
vi. Collection of seed and mycorrhizal fungi for propagation/research
vii. Promoting community awareness
viii. Obtaining biological and ecological information of surrounding habitat
ix. Research effective translocation/reintroduction strategies
x. Map habitat critical to species survival

3. Implementation
   a. Identify local, state, and federal commitments to the recovery plan
   b. Identify action officers who are responsible for completing specified recovery actions
   c. Duration and sunsetting of the recovery plan

2.6 INTERNATIONAL CONSERVATION INSTRUMENTS

The conservation status of plants and animals is one of the most widely used indicators for assessing the condition of ecosystems and their biodiversity. It also provides an important tool for establishing priorities for species conservation. At the global scale, the peak source of information on the conservation status of plants and animals is the IUCN Red List of Threatened Species (see www.iucnredlist.org; (IUCN 1999)). The Red List is designed to determine the relative risk of extinction, with the main purpose of cataloging and highlighting those taxa that are facing a higher risk of extinction. It provides taxonomic, conservation status, and distribution information on taxa that have been evaluated using the IUCN Red List Categories and Criteria: Version 3.1 (IUCN 2001).

Orchids present unique challenges when they are evaluated for Red Listing (the *gold standard* in global recognition of plant rarity). A key aspect for Red Listing species, even obviously rare species, is the need for regular monitoring over time at the population or subpopulation level to show continued decline or lack of population security. Since the accurate listing of a species requires robustly collected data on the area occupied by the species and the number of reproductive plants at each location to calculate the percentage decline and other measures, a lack of quantitative data can hinder meaningful assessment of the status of a species. It may also be necessary to document the specific threats to each subpopulation. While the subpopulation level data for orchid taxa are likely to be better organized within countries or states, this is not the case when evaluating taxa with range extensions into other countries (e.g, *Cypripedium calceolus* in Europe and the United Kingdom). Despite these challenges, the assessment process for the IUCN listing allows compilation of available data, clarifies taxonomic ambiguities, and identifies conservation critical knowledge and data gaps.

2.6.1 How the Red Listing Process Works

There are nine categories, ranging from Least Concern for species that are not threatened, to Extinct. The IUCN Red List Categories are based on a set of quantitative criteria linked to population trends; population size and structure; and geographic range. Species classified as Vulnerable, Endangered, and Critically Endangered are considered as *threatened*. When conducting regional or national assessments, two additional categories are used: Regionally Extinct and Not Applicable for non-native species (see Figure 2.11). The IUCN Red List is intended to be policy-relevant, and can be used to inform conservation planning and priority setting processes; but it is not intended to be policy-prescriptive and is not, in itself, a system for setting biodiversity conservation priorities.

The IUCN Red List Categories and Criteria method is intended to be an easily and widely understood system for classifying species at high risk of global extinction. The general aim of the system is to provide an explicit, objective framework for the classification of the broadest range of species according to their extinction risk.

The IUCN Red List Categories and Criteria have the specific aims to:

- Provide a system that can be applied consistently across global jurisdictions
- Improve objectivity by providing users with clear guidance on how to evaluate different factors that affect the risk of extinction
- Provide a system that provides a common platform for conservation assessment globally

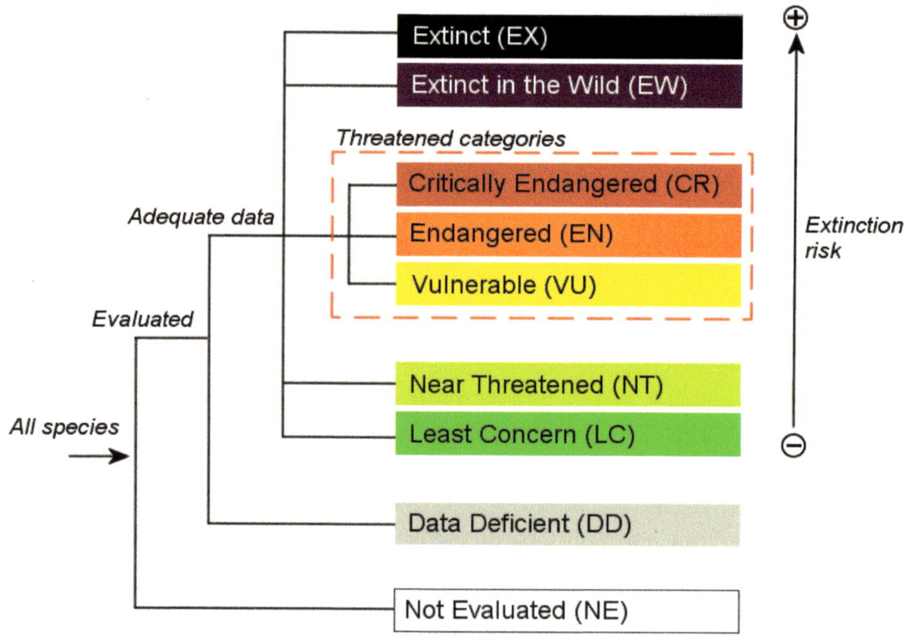

Figure 2.11 IUCN Red List categories http://www.iucnredlist.org/

3

ORCHID MYCORRHIZAL ASSOCIATIONS

3.1 BACKGROUND

One of the most significant ecological interactions orchids have with the environment is their association with a mycorrhizal fungus to facilitate seed germination, growth, and development of the plant. This functional relationship underlies many of the unique traits that are characteristic of the Orchidaceae, such as microspermy, aggregated pollen, often in pollinia, and production of often achlorophyllous protocorms. Mycorrhizal associations are a key feature of even the most phylogenetically basal orchids and are now a recognized synapomorphy for the family (Stern and Warcup 1994).

Mycorrhiza (fungus-root) are symbiotic associations between plants and fungi and are involved in the transfer of nutrients for plant growth and survival (Brundrett 2004) and germination in orchids and other plants such as certain heterotrophic lobelioids. The plant-fungus association involves direct contact between cells within the plant root (or other plant organs in contact with the substrate) and hyphae. Between 80–90% of plants globally form mycorrhizal associations. While most mycorrhiza are considered mutualistic or *balanced* where both organisms reciprocate substantial nutritional benefits to each other, there are a number of different types of mycorrhiza that involve exploitative or antagonistic associations. Orchid mycorrhiza are phylogenetically distinctive from other soil fungi groups and differ morphologically in their colonization patterns and culture on nutrient agar petri dishes (Currah et al. 1997; Hadley 1982; Rasmussen 1995; Smith and Read 1997). For further reading on other mycorrhizal systems, see Brundrett (2004).

The role of the orchid mycorrhizal association in the mature plant will be the focus of this chapter, while its role in seed germination and orchid propagation will be discussed in Chapters 4, 5, and 6. Please refer to Table 3.1 for definitions of terms used in this chapter.

3.2 ORCHID MYCORRHIZAL DEPENDENCY AND NUTRITION

The Orchidaceae display a continuum of differing levels of dependency on mycorrhiza, extending from fully-autotrophic to fully-heterotrophic orchids, based on exploitative/parasitic associations with fungi. For many orchid species the nature of their mycorrhizal

Table 3.1 Definitions of terms used in this chapter

Term	Definition
Mycorrhiza	A symbiotic association between a fungus and plant roots. Generally, the association is mutualistic, however, it can also be pathogenic.
Symbiosis	A relationship between two different organisms that produces some benefit to one or both organisms.
Endophyte	An organism, usually a fungus or a bacteria, that lives inside a plant in a mutualistic or parasitic association.
Peloton	Dense coils of fungal hyphae living within orchid roots, underground stems, and protocorms.
Mycoheterotroph	Orchids that lack or are deficient in chlorophyll for some or all of their life cycle and depend entirely on their mycorrhizal partner for nutritional support.
Saprophyte	A fungus that lives on dead or decaying organic matter.
Parasite	A plant or animal that lives in or on a host (another animal or plant, or in the case of orchids—a mycorrhizal fungus) where there is no benefit to the host—in the case of orchids, it is the orchid plant that potentially parasitizes the fungus.
Specificity	The number of partner species in a given interaction. For instance, an orchid may demonstrate high mycorrhizal specificity if it associates with only one or very few fungal partners.
Colonization/infection	The intercellular presence of fungal pelotons in underground structures of orchid plants.
Basidiomycetes	Fungi that produce spores that are formed outside a pedestal-like structure, the basidium. The members of this phylum include all the fungi with gills or pores, including the familiar mushrooms and bracket fungi. They also include key orchid-mycorrhizal fungi genera such as *Sebacina, Epulorhiza*, and *Tulasnella*.
Rhizoctonia	A genus of plant pathogenic fungus with a wide host range and worldwide distribution. Many orchid-mycorrhizal fungi were originally classified under this genus name.
Efficacious	Capability of producing a desired effect, for example, the ability of mycorrhizal fungus to initiate seed germination.
Koch's postulates	Criteria used to establish a causal relationship between a causative agent and a disease. Can be used to test the functional significance of an orchid-mycorrhizal association.
Ecto-mycorrhiza	A type of mycorrhiza composed of a fungus sheath around the outside of the root tips, with individual hyphae penetrating between the cortical cells of the root to absorb photosynthates.
Monophagous	An association where orchids receive nutrition from only one type/species of mycorrhizal fungus.

associations has not been investigated fully and only assumptions can be made on their dependence on fungi for nutrition.

Orchid mycorrhiza differ from typical ectomycorrhizal or arbuscular mycorrhizal associations as they provide a carbon source in addition to the supply of mineral nutrients to their orchid hosts (Rasmussen 1995). The carbon compounds can come from the breakdown of organic material in the surrounding substrate, exudates from adjacent non-orchid roots, and also direct transfer from an autotrophic (green) plant to the orchid. Recent broad-scale screening (Gebauer et al. 2016) of the level of mycorrhizal dependency in orchids using isotopic discrimination approaches has shown in terrestrial species, a continuum from minor to major dependency by the orchid for a fungal-derived nutrition (see Case Study 3.1). Radioactive carbon has been used to trace carbon movement from fungi to plants but not from plants to fungi; however, new research has shown that the movement of nutrients may indeed be two-way (Fochi et al. 2016). The plant may supply essential vitamins or amino acids to fungi in some cases (Leake 1994). There is no definitive evidence that fungi receive substantial benefits from associating with orchids, as they can grow independently of the host orchid (Rasmussen and Rasmussen 2009).

Case Study 3.1

Stable isotope natural abundance analysis as a tool for understanding orchid mycorrhizal nutrition

by Gerhard Gebauer and Julienne M-I. Schiebold
University of Bayreuth, Germany, Laboratory of Isotope Biogeochemistry

Most elements of biological interest are composed of two or more stable isotopes. Isotopes of an element have identical numbers of protons and electrons. However, they are distinguished by the number of neutrons in their nucleus and therefore have different atomic mass units. These differences in atomic properties of isotopes cause slight thermodynamic and kinetic isotope effects that change the isotopic composition of ecosystem compartments in predictable ways as elements cycle through the biosphere. A characteristic feature of these isotope effects is the enrichment of heavy isotopes along food chains (DeNiro and Epstein 1978), which cause fungi as heterotrophic organisms being enriched in heavy isotopes of nitrogen (Gebauer and Dietrich 1993), carbon (Gleixner et al. 1993), and hydrogen (Ziegler 1995), in comparison to accompanying autotrophic plants.

Due to their tiny, endospermless seeds, all orchids rely for germination and development in the early seedling (protocorm) phase on full nutritional support by suited fungi and are, like fungi, enriched in heavy isotopes. The degree of protocorm heavy isotope enrichment depends on the identity of their fungal hosts. Orchids germinating with fungi that simultaneously form ectomycorrhizas with forest trees are more enriched in the heavy carbon and nitrogen isotopes, ^{13}C and ^{15}N, than orchids germinating with fungi of the polyphyletic rhizoctonia group (Stöckel et al. 2014). Later on, the majority of orchids develop green leaves and become putatively autotrophic, yet remain in various states of mycorrhizal dependency. However, about 1% of all orchid species never develop green leaves and depend on fungal carbon and mineral nutrient support throughout their life cycle. These orchids, known as full mycoheterotrophs (Leake 1994), have been identified as heavily enriched in ^{13}C and ^{15}N in comparison to accompanying autotrophic reference plants (Gebauer and Meyer 2003; Trudell et al. 2003). Later research found that different patterns in ^{13}C and ^{15}N enrichment of fully mycoheterotrophic orchids depended upon the type of fungal hosts (Lee et al. 2015b; Ogura-Tsujita et al. 2009). These different patterns enable distinction between fully mycoheterotrophic orchids associated with ectomycorrhizal fungi or saprotrophic wood- or litter-decaying fungi (Figure 3.1a). Examples of fully mycoheterotrophic orchids are *Epipogium aphyllum*, *Gastrodia fontinalis*, and *Galeola falconeri* (Figure 3.2a–c).

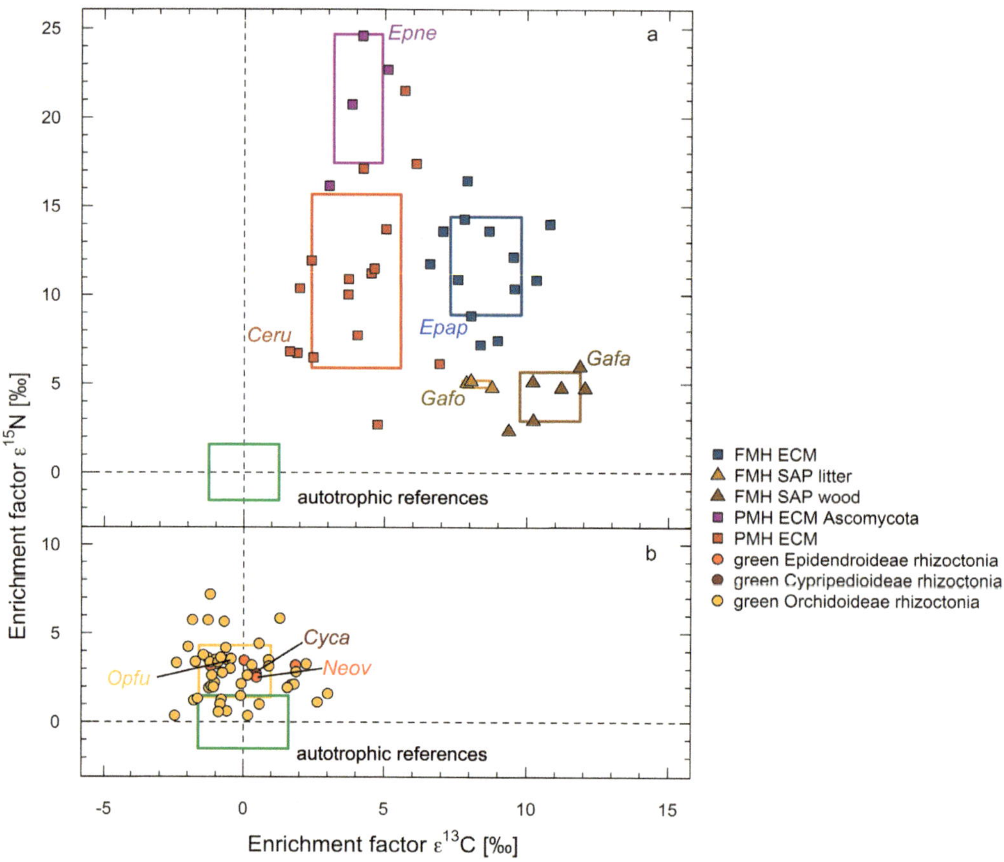

Figure 3.1 Mean enrichment factors $\varepsilon^{13}C$ and $\varepsilon^{15}N$ of fully (FMH) and partially (PMH) mycoheterotrophic Orchidaceae species (for definition of the enrichment factor ε see Preiss and Gebauer (2008)). The boxes represent ± 1 SD of the mean ε values for the different groups of PMH, FMH, and green Orchidaceae species and autotrophic references. (a) FMH orchid species associated with ectomycorrhizal fungi (FMH ECM) (n = 13; 135 data points with n = 3-62 per species), FMH orchid species associated with saprotrophic litter-decomposing fungi (FMH SAP litter) (n = 3; 15 data points with n = 5 per species), FMH orchid species associated with saprotrophic wood-decomposing fungi (FMH SAP wood) (n = 6; 31 data points with n = 1-10 per species), PMH Orchidaceae associated with ectomycorrhizal Ascomycota (PMH ECM Ascomycota) (n = 4, 28 data points with n = 5-9 per species), PMH Orchidaceae associated with ectomycorrhizal fungi (PMH ECM) (n = 16, 285 data points with n = 2-72 per species) and autotrophic reference species (n = 197, 1459 data points). (b) Green-leaved Orchidaceae species from the subfamilies Epidendroideae (n = 4, 42 data points with n = 5-30 per species), Cypripedioideae (n = 1; *Cypripedium calceolus* with n = 5) and Orchidoideae (n = 45; 405 data points with n = 4-49 per species) and autotrophic reference species (n = 162, 1302 data points). One orchid species is illustrative for each of the eight groups; photographs are shown in Fig. 3.3. Abbreviations are as follows: *Epap* = *Epipogium aphyllum*, *Gafo* = *Gastrodia fontinalis*, *Gafa* = *Galeola falconeri*, *Epne* = *Epipactis neglecta*, *Ceru* = *Cephalanthera rubra*, *Neov* = *Neottia ovata*, *Cyca* = *Cypripedium calceolus*, *Opfu* = *Ophrys fuciflora*. Data extracted from Yagame et al. (2012), Hynson et al. (2013b), Roy et al. (2013), Johansson et al. (2015), Stöckel et al. (2014), Ercole et al. (2015), Lee et al. (2015a), Liebel et al. (2015), Gebauer et al. (2016), Hynson et al. (2016).

Figure 3.2 (a) *Epipogium aphyllum*, Norway spruce (*Picea abies*) forest, Nord-Trøndelag, Central Norway. (b) *Gastrodia fontinalis*, dense bamboo (*Phyllostachys edulis*) forest, Nantou County, Taiwan. (c) *Galeola falconeri*, broadleaf forest dominated by *Phoebe formosana* and *Machilus japonica* (Lauraceae), Nantou County, Taiwan. (d) *Epipactis neglecta*, broad-leaf forest dominated by beech (*Fagus sylvatica*), NE Bavaria, Germany. (e) *Cephalanthera rubra*, mixed deciduous and coniferous forest dominated by Norway spruce (*Picea abies*), Vorarlberg, Austria. (f) *Neottia ovata*, coniferous forest dominated by Norway spruce (*Picea abies*), S Bavaria, Germany. (g) *Cypripedium calceolus*, broad-leaf forest dominated by beech (*Fagus sylvatica*), NE Bavaria, Germany. (h) *Ophrys fuciflora*, open grassland, N Italy (photos: Heiko Liebel (a) & (h), Yung-I Lee (b) & (c), Florian Fraaß (d), Gerhard Gebauer (e) and Julienne Schiebold (f) & (g)).

In adult green orchids, some species have been identified as being positioned between autotrophic reference plants and fully mycoheterotrophic plants in their carbon isotope composition (Gebauer and Meyer 2003) and associate with ectomycorrhizal fungi (Bidartondo et al. 2004). These orchids, known as partial mycoheterotrophs, gain carbon from two sources—photosynthesis and fungal hosts. Recent investigations elucidated a further separation of these partially mycoheterotrophic orchids into two groups based on their nitrogen isotope composition. Partially mycoheterotrophic orchids, such as *Epipactis neglecta* (Figure 3.2d), exclusively associate with ectomycorrhizal Ascomycota and are significantly more enriched in ^{15}N than partially mycoheterotrophic orchids associated with ectomycorrhizal Basidiomycota (Figure 3.1a), like *Cephalanthera rubra* (Figure 3.2e). Full and partial mycoheterotrophy allows orchid's thriving in the deepest shade of forest grounds or even below ground (like the Western Australian *Rhizanthella gardneri*).

In contrast, the isotopic composition of the large group of green orchids mycorrhizal with fungi of the polyphyletic rhizoctonia group remained puzzling. Irrespective of whether belonging to the orchid subfamilies Epidendroideae (like *Neottia ovata*, Figure 3.2f), Cypripedioideae (e.g., *Cypripedium calceolus*, Figure 3.2g), or Orchidoideae (e.g., *Ophrys fuciflora*, Figure 3.2h), these orchids as a group are not enriched in ^{13}C, but are significantly enriched in ^{15}N, in comparison to autotrophic reference plants (Figure 3.1b). Hynson et al. (2013a) speculated whether a kind of *cryptic partial mycoheterotrophy* might account for the isotope abundance pattern of rhizoctonia mycorrhizal orchids. In a recent investigation by Gebauer et al. (2016), adding hydrogen stable isotope abundance to the pattern of carbon and nitrogen isotope abundance confirms this assumption. This study found that seven green forest orchid species, irrespective of whether associated with ectomycorrhizal or rhizoctonia fungi and irrespective of whether

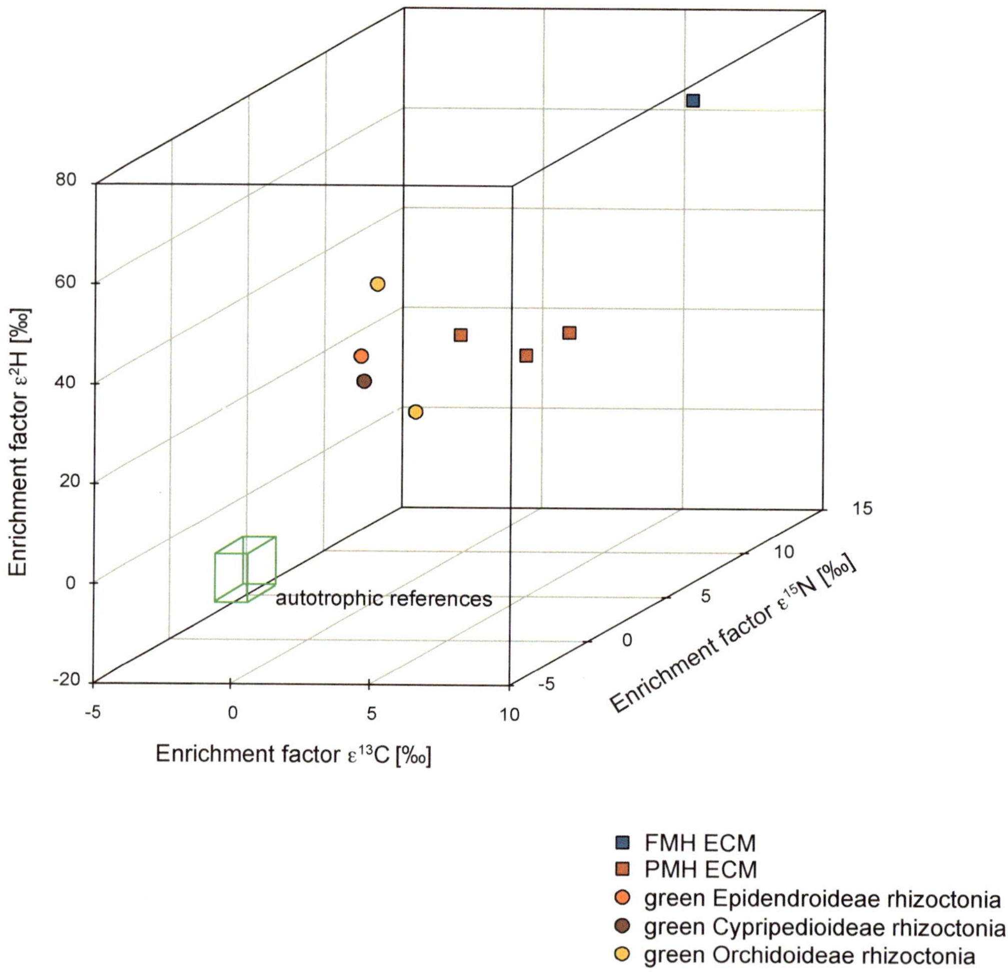

Figure 3.3 Mean enrichment factors $\varepsilon^{13}C$, $\varepsilon^{15}N$, and $\varepsilon^{2}H$ of fully mycoheterotrophic Orchidaceae species associated with ectomycorrhizal fungi (FMH ECM) (n = 1; *Neottia nidus-avis* with n = 5), partially mycoheterotrophic Orchidaceae species associated with ectomycorrhizal fungi (PMH ECM) (n = 3; *Cephalanthera damasonium* and *C. rubra*, *Epipactis atrorubens;* 16 data points with n = 5-6 per species), green-leaved Orchidaceae species from the subfamilies Epidendroideae (n = 1; *Neottia ovata* with n = 5), Cypripedoideae (n = 1; *Cypripedium calceolus* with n = 5), and Orchidoideae (n = 2; *Ophrys insectifera* and *Platanthera bifolia* with n = 5 per species) and autotrophic reference species (n = 10, 146 data points) sampled in a broad-leaf forest dominated by beech (*Fagus sylvatica*), NE Bavaria, Germany. The box represents ± 1 SD of the mean ε values for the autotrophic references. Data extracted from Gebauer et al. (2016).

enriched in ^{13}C or not, were significantly enriched in the heavy hydrogen isotope ^{2}H and therefore, were positioned between autotrophic references and a fully mycoheterotrophic orchid (Figure 3.3). This finding supports the conclusion that it is likely that many more orchids than previously assumed follow the partially mycoheterotrophic mode of nutrition.

While in most plants, the development of parasitism on its fungal partner can lead to an irreversible loss of photosynthetic capacity, in orchids, this trend does not hold. Most of the 25,000 species of orchids (Dressler 2005) have retained photosynthetic capacity while requiring a mycorrhizal fungus for seed germination and seedling development. Leake (1994) suggested that only about 170 species are fully mycoheterotrophic or achlorophyllous for their entire life history. In terrestrial orchids, early stages of germination involve a mycoheterotrophic stage followed by mixotrophy, as plants with green leaves receive nutrients through photo-assimilation, supplemented by nutrients and carbon supplied by the mycorrhizal partner (Rasmussen and Rasmussen 2009) and root-sourced soil or substrate nutrients. The transitional physiology between mycotrophy and phototrophy is poorly understood; however, it is thought that mature orchids with green leaves are substantially less reliant on mycotrophic supplementation.

In herbaceous terrestrial orchids, their few, usually unbranched roots (some orchids have mycorrhiza only in vertical underground stems, such as the Australasian genera *Pterostylis, Caladenia,* and *Eriochilus*—see Figure 3.4), indicate a predisposition of the plant for an ongoing mycorrhizal dependency (Rasmussen and Whigham 2002). In addition, colonization by mycorrhizal fungi in mature photosynthetic epiphytic orchids is often sporadic, indicating less reliance on mycorrhiza than is found for herbaceous terrestrial taxa (Rasmussen 1995).

All orchids that have been studied form endogenous mycorrhizal structures colonizing the cortical cells (Figure 3.5) of the roots, stem, rhizomes, or protocorms as densely packed hyphal coils or pelotons (Harley and Smith 1983). There is considerable variation in the distribution of mycorrhizal colonization of orchid roots and stems, and distinct colonization patterns are often associated with different orchid genera (Figure 3.6) (see Ramsay et al. 1986 for more details). Ramsay et al. (1986) also showed that different colonization patterns of the roots, collars, stems, or rhizomes can be associated with particular fungal groups; and Brundrett (2004) showed that morphological features/structures can be correlated with the presence of certain fungi.

Figure 3.4 (a) *Pterostylis;* (b) *Caladenia;* and (c) *Eriochilus* (photos: Eric Swarts)

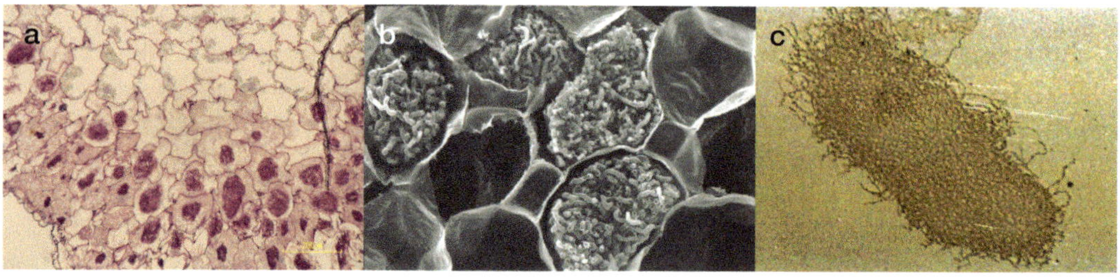

Figure 3.5 Orchid-mycorrhizal pelotons. (a) Cross section of a *Caladenia huegelii* stem (collar) illustrating mycorrhizal colonization in the outer cortical cells (×4 magnification). Scale bar = 1 mm. (b) Scanning electron micrograph of pelotons within cortical cells of orchid root tissue. Scale bar = 0.25 mm. (c) Individual peloton containing masses of undifferentiated hyphae shown under a compound microscope at 40×. Scale bar = 200 μm.

Underground stem-collar	Root tubers	Underground stem	Root	Stem tuber/rhizome
Colonization occurs within a heavily swollen region 3–10 mm long situated at or just below the soil surface subtending the basal leaf.	Adventitious root tubers arising from an underground stem or rhizome are usually heavily colonized.	Colonization occurs within a 10-50 mm section of the vertical underground stem between the soil surface and the tuber with highest density of infection contained within the upper part decreasing downwards.	This category comprises the vast majority and widest variety of genera. Adventitious roots are infected at random intervals along their length.	Mycorrhizal fungus is confined to localized patches of the outer cortex with aging of the fungus away from the apex.
Caladenia, Cyanicula, Elythranthera Drakaea	*Diuris, Cryptostylis Neottia, Dipodium*	*Pterostylis, Eriochilus*	*Ophrys, Cephalanthera, Calochilus, Thelymitra, Prasophyllum, Microtis, Dactolorhiza*	*Rhizanthella, Gastrodia*

Figure 3.6 Mycorrhizal colonization patterns in terrestrial orchids. Epiphytic orchids not shown as only roots are colonized—often at low or nil frequency along the root length. In all orchids, mycorrhizal colonization in wild germinated seedlings is most intensive in the early seedling (protocorm) stage.

3.3 IDENTITY AND SPECIFICITY

Research into orchid mycorrhizal identity and specificity varies from single peloton isolations and culture growth (Ramsay et al. 1986) to complex *in vitro* cross-specificity germination experiments (Bonnardeaux et al. 2007) and fungal *baiting* assays (Brundrett et al.

2003). These methods, used in conjunction with molecular data (Shefferson et al. 2005; Taylor et al. 2003), particularly phylogenetics, permit accurate inference of the nature and breadth/specificity of the orchid mycorrhizal association. Mycorrhizal fungal identification has traditionally been assessed by morphological comparison of cultures isolated from infected tissues of mature plants, with a variety of fungal growth and morphological criteria used for identification. Though accurate to a degree, relying solely on non-molecular approaches may result in an inconsistent taxonomy (Moore 1987).

Orchid mycorrhizal fungi are usually basidiomycetes, from the form-genus *Rhizoctonia* (Currah et al. 1997; Sneh et al. 1991). The consistent feature of orchid mycorrhizal mycelium in culture is the formation of loose aggregates of hyphae which appear as poorly developed sclerotia or resting bodies (Hadley 1982) and in some cases, the formation of bead-like structures known as monilioid cells (Figure 3.7). Some fungal isolates have been coaxed into production of sexual stages in culture, however achieving sexual expression is a difficult and unreliable procedure (Warcup and Talbot 1967; Warcup and Talbot 1971) and is rarely used in mycorrhizal taxonomy.

Fungi forming orchid mycorrhiza, confirmed through both traditional morphological and molecular approaches, belong to the basidiomycetous genera *Ceratobasidium, Ceratorhiza, Epulorhiza, Sebacina, Thanatephorus*, and *Tulasnella* (Currah et al. 1997; Warcup and Talbot 1971; Warcup and Talbot 1980). High resolution electron microscopy has also identified ascomycetous fungi forming pelotons within roots of *Epipactis microphylla* (Selosse et al. 2004). Molecular data confirmed that some species of the ascomycete genus *Tuber* and other ascomycete fungi colonized the majority of *Epipactis* root samples investigated (Bidartondo et al. 2004; Selosse et al. 2004). A wide variety of other fungi have also been isolated/identified from orchid roots, but these have not yet proven to be functionally beneficial to the orchid (Currah et al. 1997; Dearnaley 2007).

Interestingly, the form-genus *Rhizoctonia* contains saprophytes, pathogens, micro-parasites and also orchid symbionts, and the degree of cross over between members of these different subgroups is unclear. Orchid fungi are considered to be functionally differentiated from other types of mycorrhizal fungi as they typically have dual roles, both as saprophytes and parasites. In addition, given that mycorrhizal fungi appear not to receive obvious benefits from the orchid, it is unlikely that orchid fungi would co-evolve with their hosts. Thus, orchid fungi have disparate lineages from other fungal groups and the recruitment of new fungal lineages by orchids probably continues today (Batty et al. 2002).

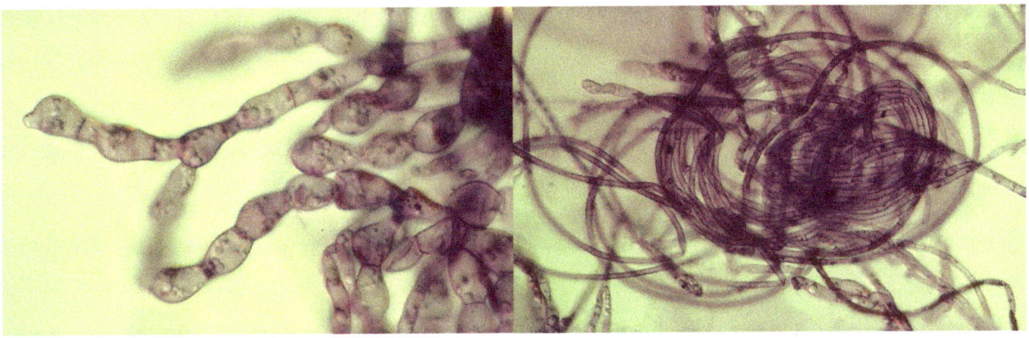

Figure 3.7 Monilioid cells of a Basidiomycete fungus. Hyphae of *Tulasnella deliquescens* (*calospora* sensu Warcup and Talbot), stained in Toluidine Blue O (photo: Hanne Rasmussen).

Specificity between mycorrhizal type and orchid species has been a topic of considerable research. In his study of tropical epiphytic species, Lewis Knudson demonstrated that orchids exhibited low levels of specificity with their fungal associates (Knudson 1927). Conversely, Burgeff (1909; 1959), who worked with mainly terrestrial taxa, reported strong orchid mycorrhizal specificity in the association and described seven categories of seed/fungus relations which ranged from fully compatible to fully incompatible. Additional evidence available today suggests that both scenarios apply (Arditti 1992).

Recent research has revealed high specificity in some terrestrial orchids (Swarts et al. 2010), while others have demonstrated that other orchid species have a wide variety of fungal associates spanning different fungal families (McCormick et al. 2004). Culture and germination tests have demonstrated that there may be many races or strains, varying in physiological function and morphology, within a species or a group of similar isolates (Phillips et al. 2016). Yet other studies have demonstrated evidence of a *super fungus* where the isolate is capable of germinating seed from a very wide range of species tested (see Case Study 3.2).

Case Study 3.2

Not all orchid mycorrhiza are created equal: evidence for 'super fungi'

by Jason Downing
Fairchild Tropical Botanic Garden, Coral Gables, FL, USA

Understanding the diverse fungal community on which orchids rely is critical to orchid conservation because the presence of orchid mycorrhizal fungi (OMF) in the environment can determine where, and in what abundance, orchids occur (Jacquemyn et al. 2014; McCormick et al. 2012; Swarts et al. 2010). Identifying these specific fungal taxa should thus be prioritized in orchid conservation studies. Of particular interest are OMF that form beneficial associations with more than one orchid species. These *super fungi* are often capable of associating with multiple orchid species that can span genera, distributional ranges, and life histories (Zettler et al. 2013).

One such *super fungus* has recently been discovered in tropical China. In Guangxi province, a distinct OMF was detected in the roots of *Paphiopedilum hirsutissimum*, *Paphiopedilum dianthum*, and *Cymbidium bicolor* (Figure 3.8). These orchid species represent both wide and narrow latitudinal range distributions. The same fungus has also been detected in *Paphiopedilum micranthum*, *Cymbidium lancifolium*, and *Holcoglossum sinicum* collected in the same region (Tan et al. 2012; Yuan et al. 2010).

Another *super fungus* grows in southern Florida, and may help explain some recent plant invasions. An OMF was cultured from the protocorms of two invasive terrestrial orchid species (*Cyrtopodium flavum* and *Eulophia graminae*). This fungus was also capable of germinating the seeds of three native species: *C. punctatum*, *E. alta*, and *Triphora gentianoides*—with *C. punctatum* being an extremely rare epiphytic species (Wunderlin and Hansen 2004). Also in southern Florida, the fungus *Tulasnella irregularis* (Basidiomycota: Tulasnellaceae) was cultured from roots of *Encyclia tampensis* and subsequently induced germination in three sympatric species, *Epidendrum amphistomum*, *Epidendrum nocturnum*, and *Prosthechea cochleata*.

Figure 3.8 (a) *Cyrtopodium flavum* adult invaded Pine rocklands in Miami, Florida. (b) *Cyrtopodium flavum* protocorms (germinated with *super fungi*). (c) *Paphiopedilum hirsutissimum*. (d) Southern Florida *super fungus* (photos: Jason Downing).

The broad phylogenetic compatibility of these (likely) keystone OMF demands further research to identify their full life history and environmental requirements. Additionally, *ex situ* preservation and management of the habitats that support these *super fungi* are of crucial conservation importance.

For epiphytic orchids, although the number of studies is fewer, the fungal association is considered to be broad and less dependent, both at the seedling and adult stage. Yet, Otero et al. (2007) demonstrated a lifelong association in the neotropical epiphyte *Ionopisis utricularioides*. Clearly though, the industrial-scale commercial production of nonmycorrhizal epiphytic orchids (*Phalaenopsis, Cymbidium, Oncidium, Paphiopedilum*) points to the mycorrhizal dependency in epiphytes being far less stringent than for herbaceous terrestrial taxa.

It has been reported that host-fungus specificity is generally lower at the time of germination than it is during seedling development *in vitro* (Muir 1989) and then appears to decrease again as the plant reaches the photoautotrophic phase, when other fungi may be encountered (Warcup 1981; Alexander and Hadley 1983). Thus, the fungus associated with seed germination may not remain with the orchid plant, as found for *Gastrodia* (Xu and Guo 2000), although there are many examples where fungi isolated from adult orchids promote successful germination (see Arditti 1992 for examples). Ramsay et al. (1987) used hyphal anastomosis groups (AGs) to categorize fungi associated with the genus *Pterostylis* and showed that fungal types may be linked to particular habitat types. For example, AG8 is only recorded for warm, arid habitats; AG3 in semi-arid sites (e.g., exposed rock habitats); and AG4 is only associated with *Pterostylis* species that grow in introduced pine plantations. Conversely, vicarious fungal groups such as AG2 isolates, associate with *Pterostylis* species across a range of habitats. Sympatric species of *Pterostylis* have also been found to use different AG's. Different fungal types have been shown to associate specifically with genera of Western Australian orchids (Ramsay et al. 1986; Ramsay et al. 1987). For example, in sterile culture on potato dextrose agar (PDA), all *Caladenia* species examined use a mucoid, slow growing, cream-colored fungus, whereas *Pterostylis* species require a fast growing, fluffy, white culture for *in vitro* seed germination and seedling growth. Though molecular methods can provide more definitive characterization of fungi, techniques such as AG associations are a useful tool in understanding the degree to which fungi may *share* common hyphal networks in soil.

It appears that the specificity, dependency, and diversity of the mycorrhizal association in orchids can be attributed to the underground structures in which the fungi are associated. A single point underground root stem-collar infection, as seen in many Australian terrestrial orchids, (e.g., Caladeniniae) appears to almost always relate to a highly specific, lifelong, single fungal type (Swarts et al. 2010). In this case the orchid might have associated with a highly efficacious and competitively successful fungus as its key nutritional partner; whereas, in the case of root-infected terrestrials or epiphytes with a greater photosynthetic capacity, the orchids may *forage* for a broad range of compatible fungi to supplement their nutritional requirements (i.e., mixotrophy).

Although two ends of a fungal dependency spectrum have been described here, it is most likely that a continuum exists between high dependency to low or infrequent dependency. Important in conservation, particularly in the face of climate change, a fungus with a highly specialized set of ecological requirements, although successful in its current habitat, may become significantly less efficacious and therefore at conservation risk in the event of biotic change (e.g., weeds, cover), or thermal or hydrological shifts. Indeed terrestrial orchids may have adopted highly specific mycorrhizal dependency (apart from some atypical *weedy* species such as *Microtis media* and *Disa bracteata)* (Figure 3.9) (Bonnardeaux et al. 2007).

Figure 3.9 *Microtis media* at Stage 2/3 on water agar with no added nutrients indicating that this weedy species has a capacity for germination and early growth in the absence of mycorrhiza or added nutrients.

Orchids may be able to survive in symbioses with one or many fungi, however, little is known about the importance of fungi that occur in the same location or niche. If fungi of different species can access different components of the organic/biotic nutrient pool, this may minimize competition between co-occurring orchids, if they can utilize different available fungal species.

3.4 ACHLOROPHYLLOUS ORCHIDS

A small number of orchids lack or are deficient in chlorophyll for some, or all, of their life cycle (see also Case Study 3.1). Known as myco-heterotrophs, these orchids are nourished almost entirely via their mycorrhizal partner (Leake 1994; Smith and Read 1997). The Orchidaceae has the largest number of achlorophyllous species of any plant family and these highly specialized associations have arisen independently in at least 20 independent orchid lineages (Brundrett 2002; Leake 1994; Molvray et al. 2000).

Most myco-heterotrophic plants are achlorophyllous and may occupy dark or shady environments where their ability to exploit sources of carbon unavailable to other plants may be ecologically advantageous (Brundrett 2002; Leake 1994). The loss or reduction of both leaves and roots in myco-heterotrophic, achlorophyllous orchids results in limited capacity for photosynthesis or direct uptake of water and nutrients. Consequently, these species are entirely dependent on mycotrophic support by a specific fungal partner (Bidartondo et al. 2004). This makes this group of plants one of the most vulnerable because they have a restricted distribution—occupying areas with minimal light, high humidity, and abundant leaf litter. It has been suggested that their restricted distributions result from their highly specific relationships with mycorrhizal fungi that are, in themselves, ecologically restricted (Perkins and McGee 1995; Rasmussen 1995). Consequently myco-heterotrophic orchids are rarely dominant in natural ecosystems and many species are indeed rare (Brundrett 2002). However, there are cases of some myco-heterotrophs, such as *Gastrodia,* becoming weedy when organic supplements have been used in garden situations (Figure 3.10).

Figure 3.10 Though a holomycotroph reliant on specific and obligately required mycorrhiza, *Gastrodia* has become weedy in a garden in New South Wales where the use of water-saving mulch provides the ideal germination and growth medium for this otherwise uncommon orchid to germinate and proliferate.

Figure 3.11 An extreme case of orchid nutritional dependency—the Western Australian underground orchid, *Rhizanthella gardneri*, a leafless, rootless, and fully underground growing species that relies on a complex interaction with ectomycorrhiza of the root system of a nearby shrub (various *Melaleuca* species) to provide mineral and carbon nutrition to the orchid seed.

Australia's underground orchids (*Rhizanthella* species—see Figure 3.11) lack green leaves, obvious roots, and rely on their fungal associates for nutrition. At least in the Western Australian species, the fungus is known to link to the root system of a nearby shrub species (*Melaleuca* species). Germination of the Western Australian species only occurs when the orchid seed is in direct contact with a mycorrhizal root of the host shrub, indicating one of the highest levels of ecological specialization found in orchids.

3.5 ORCHID MYCORRHIZAL METHODS

The effective use of orchid mycorrhiza depends, in large part, on an understanding of the efficacy of the mycorrhizal fungus to promote germination, growth, and development of the orchid. This often relies on isolating the mycorrhizal forming fungus from the orchid and subsequent resynthesis under axenic or soil conditions to test for germination and growth benefits. Alternative approaches, such as using direct polymerase chain reaction (PCR) to diagnose for mycorrhizal presence, have contributed significantly to defining fungal diversity (see Chapter 9) in orchid tissues. But the absence of isolation and testing for germination benefits has led to uncertainty over what constitutes an ecologically useful mycorrhizal fungus. Likewise, DNA sequencing of pelotons isolated from infected orchid tissues and cultured on nutrient agar will provide more accurate information on diversity, but again, will not determine which fungi are functionally necessary. Determination of mycorrhizal specificity must assume functionality, which can only be determined through the adoption of Koch's postulates (isolating causal agent, resynthesis *in planta*, re-isolation of the same fungus from inoculated tissues). Table 3.2 identifies techniques most commonly used in the process of determining fungal identity and diversity, functional significance, and specificity.

Table 3.2 Techniques most commonly used in the process of determining fungal identity and diversity, functional significance, and specificity.

Method	**Isolated fungal cultures (Chapter 3)***	**Fungal *baiting* in soil (Chapter 5)**	**Direct soil assay (Chapter 9)**	**Direct root assay (Chapter 9)**
Testing and purpose of method	Diversity; distributional status; functional significance.	Diversity; presence/absence; distribution status; ecological significance of fungi.	Presence/absence; quantification of fungal presence; distribution status of fungi.	Diversity; distribution status.
Description	Isolation of pelotons from infected root/stem tissue, cultured on nutrient agar followed by DNA sequencing. Fungal cultures screened for germination efficacy.	*Ex situ* or *in situ* fungal screening in collected soil or field sites. Fungal isolation and culture direct from protocorms followed by DNA sequencing. Fungal cultures screened for germination efficacy.	Direct probing of collected soil organic matter with discriminatory fungal primers using real-time quantitative PCR.	Fungal DNA extraction directly from infected plant tissues and assay using fungal specific primers.
Advantages	Easy to extract and culture pure fungi (most orchid taxa); determines a level of *potential* ecological function; provides indication of ecological specificity; accurate fungal identification; wide range of primers available.	Substantiates Koch's postulates; higher degree of confidence on functional significance; accurate fungal identification; wide range of primers available; match fungal presence or absence with soil probe data.	Capacity to screen relatively large geographical ranges; probe for specific fungal types.	Potential to assay many root samples; accurate identification of associated fungi.
Disadvantages	Multi-stage approach; cultures susceptible to contamination; distribution testing limited to orchid distribution.	Time intensive to set up for large area surveys; cultures susceptible to contamination; *in situ* method is seasonally dependant.	Costly PCR; prior knowledge of fungi required; does not determine functional significance of fungi; need to prepare highly specific fungal primers.	Costly PCR; will not determine which fungi are functionally significant; distribution testing limited to orchid distribution; require fungal specific primers.
Reference	Bonnardeaux et al. (2007); Bougoure et al. (2005); Ramsay et al. (1986); Swarts et al. (2010)	Batty et al. (2001); Brundrett et al. (2003); Jacquemyn et al. (2011); Masuhara and Katsuya (1994); Rasmussen and Whigham (1993)	Filion et al. (2003); Kennedy et al. (2007); Landeweert et al. (2005)	Shefferson et al. (2007); Taylor et al. (2002); Taylor et al. (2003)

*Modifications to this approach may be required depending on the nature and type of association, for example, ectomycorrhizal associations (ECM) where the fungus is difficult to grow in an axenic culture.

3.6 FUNGAL CULTURE ISOLATIONS

3.6.1 Plant Material Collection

In terrestrial orchids, only certain underground, non-green parts are colonized with mycorrhizal fungi. In epiphytic species, only the protocorms and roots are colonized (Figure 3.6). Colonization depends upon the amount and timing of rainfall and seasonal maturity of the plant within the growing season. The influence of seasonal duration and variation on collection time is an important consideration to ensure maximum fungal colonization in the plant material. Harvais and Raitsakas (1975) and Warcup (1973) reported that the most efficacious fungus for seed germination was isolated early in the seasonal growth cycle of the plant, i.e., at leaf emergence and prior to flowering. Pelotons isolated at later developmental stages were often moribund and contained fungal and bacterial contaminants (Huynh et al. 2004). Huynh et al. (2004) found for the rare and endangered herbaceous terrestrial species *Caladenia formosa*, the most germination-effective fungal cultures were grown from pelotons isolated from plants in the early stages of emergence and growth.

For each colonization type there are subtly different methods required to ensure a successful mycorrhizal isolation with minimal disturbance to the host plant (Table 3.3). Following excavation, orchid tissue is placed in a moist paper towel and chilled until the pelotons are extracted in the laboratory. Ideally, peloton extraction should happen on the same day as collection, but root material can be refrigerated for 3–4 days. Root material can soften and rot, with a greater chance of bacterial contamination the longer material is stored, and the pelotons become more difficult to extract.

Table 3.3 Specific methods for mycorrhizal isolation from orchids with different colonization strategies

Colonization strategy	Specific isolation methods
Root colonized orchids	Root colonized orchids are the most taxonomically diverse and geographically widespread form of orchid mycorrhizal association. In the case of root colonized terrestrial species, carefully extricate the roots from the soil. Care must be taken at this stage as orchid roots are often easily damaged. Note: many root colonized orchids have one or more *apogeotrophic roots* (roots that grow upwards or diagonally towards the surface)—these roots are often the most heavily colonized with mycorrhiza.
Stem and collar colonized orchids	Here the mycorrhizal colonization occurs in the plant just below the soil surface in what is referred to as the *collar* (as found in *Caladenia* and related taxa) or in the entire underground stem (*Pterostylis, Eriochilus*). To isolate, carefully excavate the litter layer around the plant, remove part (higher risk of contamination) or all of the swollen stem. In this way, the tuber remains intact and the plant will resprout the following growing season.
Rhizomatous/stem tuber colonized orchids	For a rhizomatous (stem tuber) colonized species, a section of the rhizome (only sample from the most recently matured section as older sections will have non-viable fungi) or, if permitted, the whole plant can be excavated and appropriate sections taken. Again, confine sampling to the most recent rhizome section, as this is where the most active fungal tissue is likely to reside.

Procedure 3.1

Clearing and staining mycorrhizal structures in orchids to estimate mycorrhizal colonization (adapted from Brundrett et al. 1996 and Batty and Brundrett 2002)

Studies of orchid mycorrhizal association in roots or stems often require estimates of the degree of colonization in tissues (number of cells occupied by fungi). Colonization of orchid mycorrhizal fungi is not usually visible when fresh roots are observed because internal structures are obscured by the natural pigments and cell contents within the roots. With the use of stains that highlight fungal tissues, these structures can be observed in plant tissues. This procedure involves a clearing stage that uses a chemical agent such as a hot alkali to remove cell contents and cell wall pigments, which then allows the staining of mycorrhizal hyphae. Fungal structures are revealed by the use of stains such as trypan blue, chlorazol black E (CBE) in lactoglycerol, or ink in vinegar (Bevege 1968; Brundrett et al. 1984; Kormanik and McGraw 1982; Phillips and Hayman 1970) which bind to fungal hyphae (Brundrett et al. 1984).

Materials:

- 50% ethanol (v/v) for tissue preservation.
- 10% potassium hydroxide (KOH) (w/v) dissolved in water.
- 0.03% w/v chlorazol black E (CBE) in lactoglycerol (1:1:1 lactic acid, glycerol, and water). Dissolve CBE in water before adding equal volumes of lactic acid and glycerol.
- 0.05% w/v trypan blue in lactoglycerol (1:1:1 lactic acid, glycerol and water).
- 50% glycerol-water (v/v) solution for destaining and storage of stained roots.
- Food grade lactic acid and glycerol are adequate and can be purchased in bulk.
- Autoclave (121°C), water bath, or oven (60–90°C) to heat samples for clearing and staining.
- Autoclave-resistant glass jars, or tubes to hold samples.
- Fine nylon mesh screen (±100 μm) to prevent root loss when changing solutions.
- Fine forceps and dissecting needles to transfer roots.
- Plastic vials with well-fitting lids for storage of stained samples in 50% glycerol.
- Clear plastic dishes with inscribed grid lines to measure colonization.
- Fine screen (100 μm) with nylon mesh for transferring roots from solutions.
- Fine forceps and probes for manipulating roots.
- Microscope slides, long cover slips.
- Plastic vials with well-fitting lids for storage of samples.
- PVLG mountant (Koske and Testier 1983).

Polyvinyl alcohol*	8.33 g
Distilled water	50 ml
Lactic acid	50 ml
Glycerine	5 ml

*Polyvinyl alcohol is dissolved in water (90°C) overnight.

Methods:

1. Place clean root/stem samples (< 2 g) in 10% KOH. KOH or staining solution volumes must be sufficient for the amount of roots being processed; make sure that the roots are not tightly clumped.
2. Autoclaving samples in glass containers using a liquids cycle of 15–20 minutes at 121°C will efficiently clear most roots. Samples containing old roots, roots with abundant phenolics, or field-collected roots often require longer clearing times (25–60 minutes).
3. Root samples can also be cleared by heating KOH in a water bath at 60–90°C. Longer times may be required (from five hours to several days).
4. Cleared samples are retrieved using a fine sieve and rinsed with water before transferring into the staining solution.
5. Cleared material is stained with CBE or trypan blue in a lactoglycerol solution.
6. Stain samples by heating for several hours in a 90°C waterbath, or by leaving them in the solution for several days. The staining solution may be reused several times, if filtered through folded cheesecloth or 50 µm nylon screen after each use (to remove plant fragments).
7. Post-clearing bleaching with alkaline hydrogen peroxide (0.5% NH_4OH and 0.5% H_2O_2 v/v in water) effectively removes excessive remaining pigments.

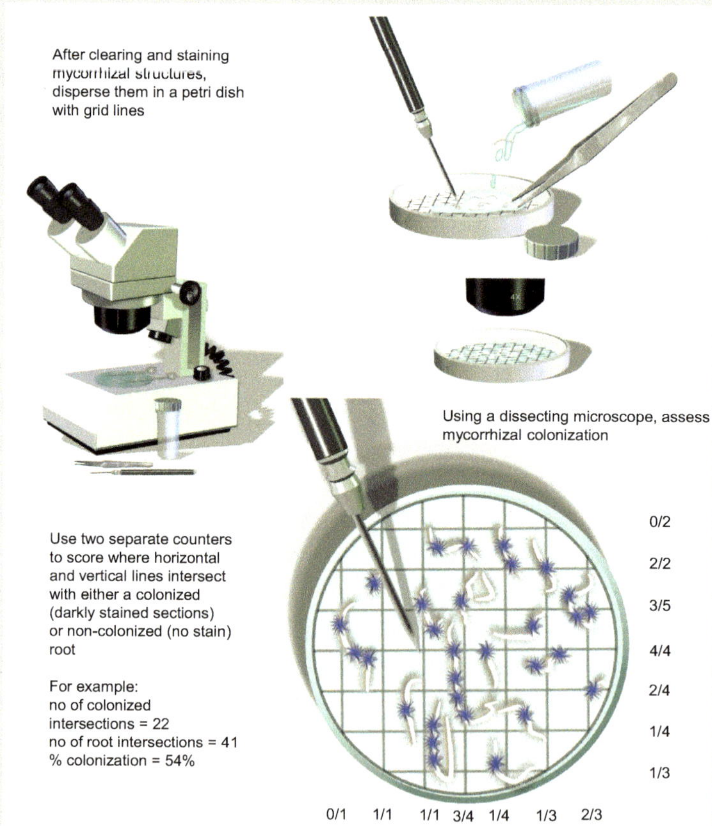

Figure 3.12 Gridline intersection method for estimating mycorrhizal colonization of orchid roots (adapted from Brundrett et al. 1996)

8. Samples should be destained in 50% glycerol (CBE) or lactoglycerol (trypan blue) for several days prior to observation to allow excess stain to leach from roots.
9. Roots stained with CBE can be stored in 50% glycerol, but trypan blue staining is less permanent, unless samples are stored in lactoglycerol or the staining solution.
10. Semipermanent slides of stained roots can be made with PVLG mountant.
11. Following staining, vital staining procedures that measure succinate dehydrogenase activity can be used to confirm that mycorrhizal fungus hyphae are metabolically active (Schaffer and Peterson 1993; Tisserant et al. 1993).
12. Complete the estimation of mycorrhizal colonization (Figure 3.12) by gridline intersection (Giovannetti and Mosse 1980; Newman 1966) or by viewing slides containing cleared samples with a compound microscope (McGonigle et al. 1990).

Safety notes

Modifications to standard clearing and staining procedures have been proposed for safety reasons.

- 10% w/v KOH is used to clear roots. A lower concentration of KOH (2.5%) can be used to reduce the risk of injury. Care should be taken to avoid skin contact with this corrosive chemical. Use gloves, safety glasses, etc.
- CBE and trypan blue are suspected to be carcinogens, as are many other biological stains. Use gloves to protect your hands when using stain solutions and take care to avoid breathing dust or getting any into your eyes when handling dye powders; and always dispense powders in a fume hood. Methyl blue or aniline blue can be used as less toxic replacements for CBE, or trypan blue, however, there is insufficient evidence to confirm that these dyes are nontoxic.
- Refer to laboratory specific safety data sheets for the latest information on the chemicals used.

3.6.2 Surface Sterilization

Following the collection of material from the field, soil and other adhering contaminants are removed by repeated washing and gentle rubbing under running tap water. Plant tissues are then dissected with a scalpel blade into small sections, focusing on those parts that contain the greatest abundance of pelotons. Locating the regions rich in pelotons is often a case of trial and error, and requires microscopic observation. If plant tissues are particularly dirty, they can be surface sterilized in a 1% sodium hypochlorite solution for 5–10 minutes, however, for root-colonized orchids, surface sterilization can result in the death of the mycorrhizal fungus of many species. Tissue block isolation requires surface sterilizing for 3–5 minutes, followed by rinsing in sterile water.

Generally, surface sterilization of orchid tissue sections following washing of adhering soil particles under running tap water requires three rinses with vigorous agitation in sterile water. This can be done in capped, 30 ml tubes that are shaken vigorously, with water changes undertaken in a laminar flow cabinet.

3.6.3 Isolation of Pelotons

Two methods are used to obtain sterile pelotons for initiation of fungal cultures. Both are completed inside a laminar flow cabinet (Figure 3.13). Either the outer layer (epidermis) of cells is excised with a scalpel blade under a dissecting microscope, and the pelotons extracted from the cortical cells, or tissue blocks (1–2 mm cubes) are cut and plated onto fungal isolation medium (FIM) or soil solution equivalent (SSE) agar. The individual peloton isolation method is recommended, as this limits the degree of contamination that can occur when using tissue blocks, as well as limiting the amount of endogenous bacterial contamination.

There are a number of approaches to isolating fungi from orchid tissues and the method described here is effective, though successful isolation depends upon a good working knowledge of sterile technique (see Text Box 3.1 and Figure 3.14).

Text Box 3.1

Use of laminar flow cabinet (Figure 3.13)

There are basic practices for use of a laminar air-flow cabinet.

1. If the laminar flow has ultraviolet (UV) capacity, turn it on 15 minutes before use. Cleanse the working surfaces before use with 70–90% ethanol. Spray all surfaces in the cabinet, including walls and ceiling, and swab down with cotton wool or a paper towel. Cleanse both before and after switching on the cabinet. Avoid breathing in the ethanol spray.
2. All items entering the cabinet should be sterile or sterilized. Wear gloves (or thoroughly wash hands) and cleanse by wiping with ethanol. Make sure glassware is clean and sterilized before placing in the cabinet. Metal instruments need to be sterilized before use by either autoclaving or dipping in 90% ethanol and flaming. The ethanol must be placed well away from the flame source.
3. Remember that any bacteria or fungal spores in the cabinet will flow down and toward you. Never place hands, sleeves, or non-sterile items above or beyond the work area (e.g., media). Keep movements smooth and avoid creating air turbulence that could bring in contamination (e.g., talking, coughing, or sneezing). Work toward the back of the cabinet as much as possible, and minimize exposure time of media where possible.
4. Maintain sterile conditions by swabbing the work surface regularly with ethanol, cleaning instruments after each use, and cleaning hands after they have been in contact with anything outside the cabinet.

Safety notes:

1. *Be careful not to flame your hands*—particularly when wearing gloves! It is easy to do as you cannot feel if the ethanol on your gloves has not totally evaporated.
2. *Be careful not to burn your hands*—the instruments are very hot.
3. *Always remember to turn off the gas*—this is particularly important even if you are leaving the laminar flow for short periods of time.
4. *Avoid inhalation of ethanol fumes.*

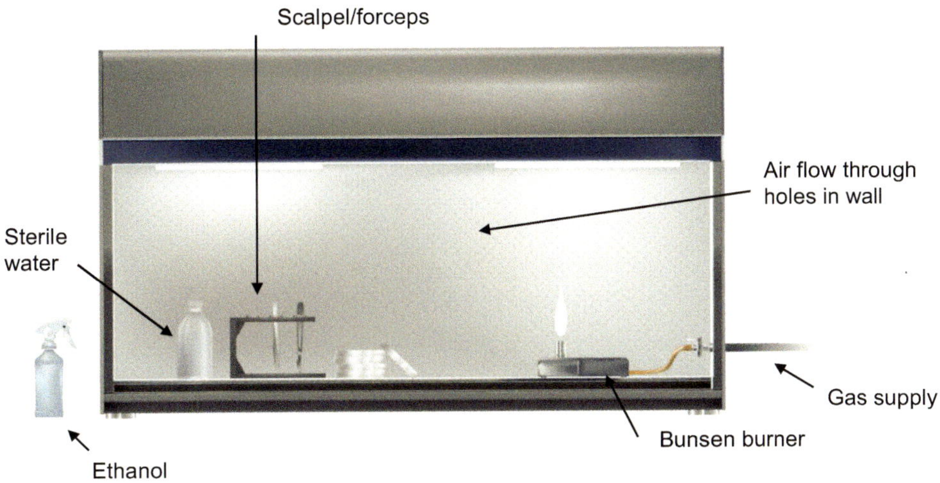

Figure 3.13 Laminar flow cabinet

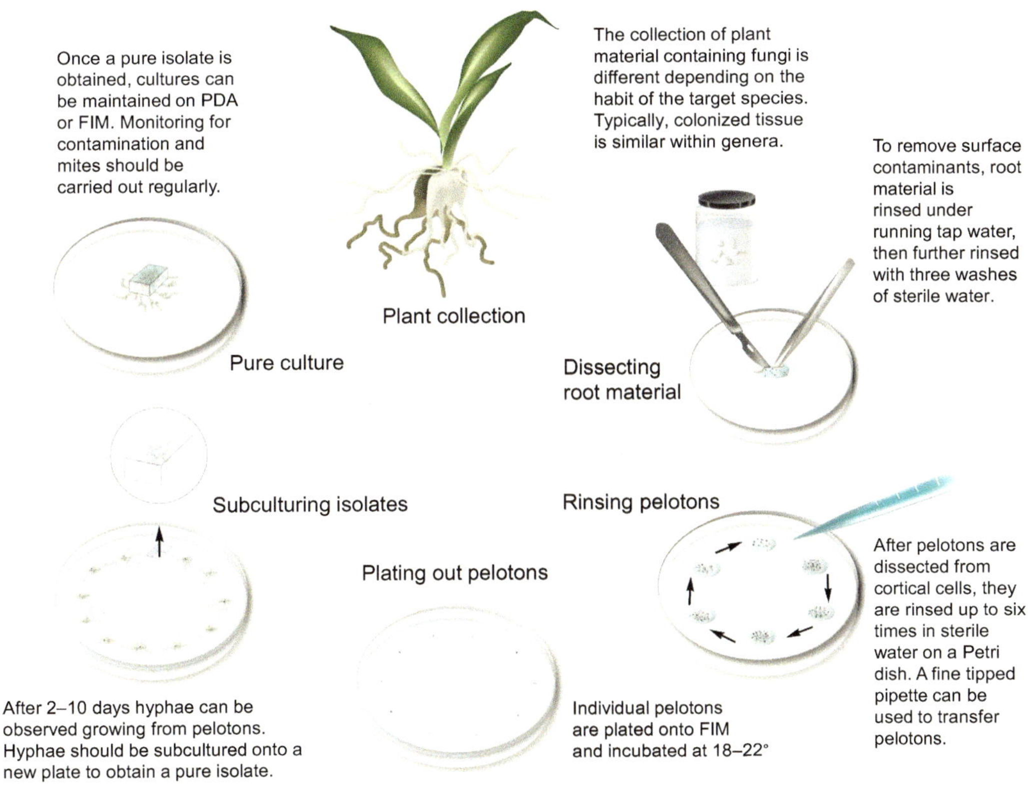

Figure 3.14 Peloton isolation protocol where colonized tissues are carefully dissected to release the small white *globules* of fungal material (pelotons). These are sequentially washed to cleanse the material of potential contaminants, then cultured on nutrient agar until pure cultures are obtained.

Procedure 3.2

Peloton isolation from orchid roots

Materials:

- Flowing tap water
- 30 ml tubes of sterile water (autoclaved for 20 minutes at 121°C)
- Bottle of sterile water
- Petri dishes containing FIM or SSE media + streptomycin
- Permanent marker
- Scalpel and forceps
- Laminar flow cabinet
- Dissecting microscope and light source
- Sterile petri dishes (empty)
- Bunsen burner
- Ethanol/methanol
- 20–200 μl pipette
- Gel loading tips
- Masking tape
- Small screw top tubes/bottles
- Latex gloves

Methods

1. *Prepare laminar flow cabinet and sterilize instruments*: Turn on the laminar flow cabinet and spray clean with ethanol. Remember, a laminar flow cabinet blows clean air from the back so obstructions to the airflow may exacerbate contamination. Ensure all ethanol has evaporated and place the tools inside the cabinet.
2. *Rinse plant tissue*: Remove orchid tissue from collection tubes. In a sink, gently wash the orchid tissue under running tap water to remove adhering debris and organic material. Place the tissue in a labeled tube containing tap water. If isolating from multiple samples, write sample details onto strips of masking tape and wrap these around each tube.
3. *Rinsing tissue in sterile water tubes*: For each tissue sample to be cleaned, three sterile water tubes are required. *Serial dilutions*: In the laminar flow cabinet, remove rinsed orchid tissue from the collection tube with forceps and place into a tube containing sterile water. Remove the masking tape label from the collection tube and wrap it around the first sterile water tube. Let each sterile rinse tube sit for approximately three minutes with gentle agitation. Resterilize the forceps, remove the orchid tissue from the rinse tube, and place the tissue into the next rinse tube. Repeat this process, ensuring that

each sample is rinsed at least three times. Do not reuse sterile water tubes for different samples.

4. *Clean the workspace*: After rinsing the tissues, clean the work surface with ethanol and wipe down with paper towels to remove contaminants. Set up the dissecting microscope and light source inside the laminar flow cabinet. Wipe the microscope with ethanol.
5. *Isolate fungal pelotons*: Place one side of an empty Petri dish under the dissecting microscope. With a gel-loading tip on a 20–200 ul pipette and using a fresh tube of sterile water, pipette three 200 µl aliquots into the middle of the Petri dish. Remove a sample of cleansed tissue and place into the water droplet. Replace the lid on the rinse tube. With sterile forceps, gently scrape the section of orchid tissue. Fungal pelotons, visible under the microscope as small, white or cream, oblong to ovoid, granular structures, should begin dispersing in the sterile water droplet.
6. *Serial dilution of pelotons*: Take the pipette and using the same tip, place another sterile water droplet to the side of the Petri dish. Collect pelotons from the isolation droplet in as little water as possible, and eject into the fresh droplet. Transfer approximately 30 pelotons at each change. Replace the pipette tip and place another sterile water droplet to the side of the Petri dish. Repeat the peloton transfer procedure 3–5 times using a fresh pipette tip between droplet changes. Some pelotons will be lost in the water changes, however, with practice, it is possible to recover 50–75% of those initially extracted.
7. *Plating pelotons*: With a fresh pipette tip, draw up individual pelotons and position in a circle around a Petri dish containing FIM or SSE medium. Try to replace the lid of the plate between each peloton placement to minimize contamination. Note: each collection tube will contain up to three pieces of orchid tissue. Each piece of tissue will undergo separate isolation, diluting, and plating. Thus, there should be three isolation plates for each species collected at a given location. This will provide a representative spread of fungal diversity and will provide enough material in the event of contamination.
8. *Dry the plate*: Place the inoculated Petri dish toward the back of the laminar flow cabinet, with the lid resting in front of the plate. Evaporating the droplet to near-dryness will minimize bacterial contamination.
9. *Repeat steps 7 and 8*: Repeat the dilution and plating process for all collection tubes, ensuring that each collection tube results in three plates, one per piece of tissue.
10. *Store plates*: Label and seal the inoculated plates with either parafilm or cling wrap and then place in an incubator at 20°C (some fungi, particularly any tropical or subtropical species, may require higher incubation temperatures).
11. *Monitor plates*: Isolation plates should be checked every two days for culture growth or contamination. Remember that some species of fungi grow more quickly than others.
12. *Isolate purification*. Microscopic examination of pelotons every few days is required until hyphae are observed growing from individual pelotons. For fast-growing isolates, hyphae are present after 2–3 days; for slow-growing fungi, hyphae may not appear for 7–10 days or months.

Bacterial contamination (Figure 3.15) can be removed by carefully subculturing outgrowths of the desired fungal material onto nutrient agar media (see Section 3.8) with the addition of antibiotics, such as streptomycin sulphate (see media preparation in upcoming text for further details). Where plates are heavily contaminated, cultures may need to be transferred to partial or window plates. These are plates that are only partially covered in agar medium. A small cube containing the contaminated isolate is placed onto the floor of the petri dish approximately 10 mm from the agar. Hyphae that grow across the floor of the petri dish are usually free of bacteria. Bacteria-free isolates can then be subcultured to nutrient agar media. Alternatively, if hyphae have grown through the agar to the base of the Petri dish, a section of the agar can be excised and flipped so that hyphae can be cultured from the bottom surface. When hyphae grow through agar they are usually free of bacterial contaminants.

To ensure genetic purity of fungal subcultures, it is important to subculture single hyphal tips from the outer edge of a colony. Although isolates have been made from single pelotons, it is not uncommon for more than one fungal type to be incorporated in a peloton. It may be necessary to complete the first subculture under a dissecting microscope to ensure the accuracy of this step. Refer to Figure 3.14 for an overview of the steps involved in obtaining pure fungal cultures.

3.7 MYCORRHIZAL CULTURE GROWTH AND STORAGE

Mycorrhizal isolation is completed using FIM or SSE media (see Section 3.8). For short-term storage (up to 1 year), fungal cultures can be maintained on plates of FIM, SSE, or potato dextrose agar (PDA) (Figure 3.16). In this way, subculturing for experimental purposes and longer term storage is made easier and more convenient. Repeated subculturing of isolates may lead to a reduction in the capacity of isolates to germinate seed.

Media in Petri dishes tend to be easily infected by mites, and condensation forming on Petri dish lids may increase the risk of contamination. For longer-term storage (1–5 years), it is recommended that fungal isolates are subcultured into screw-topped plastic tubes containing sloped PDA, and stored at 5°C. Other fungal storage protocols include storage under sterile water, in vegetable juice broth, or under oil.

Very long-term storage of fungi is best achieved using ultra-low temperatures such as liquid nitrogen (–196°C). Here isolates are protected from mite infestations or genetic change. This method is outlined in Chapter 10.

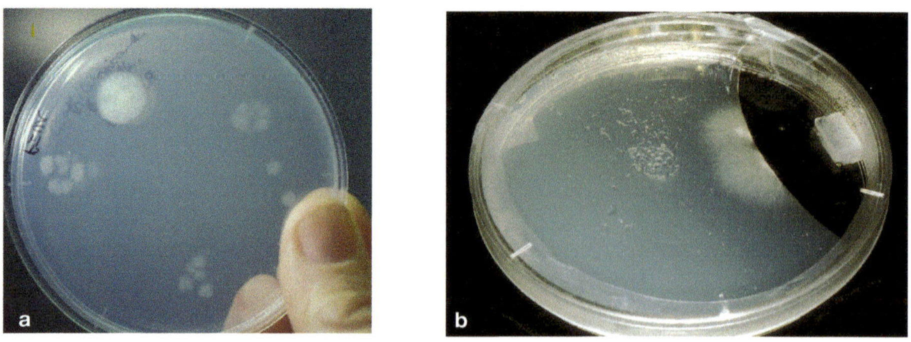

Figure 3.15 (a) Bacterial contamination on agar plate and (b) removing bacteria from fungal plates

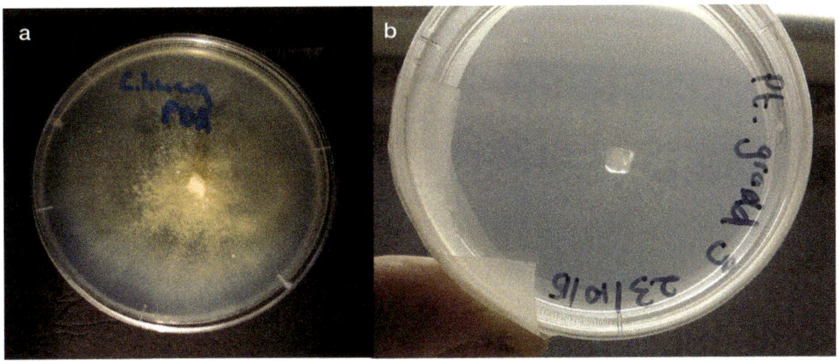

Figure 3.16 Orchid mycorrhizal cultures (a) *Sebacina* isolated from *Caladenia huegelii* on PDA and (b) *Ceratobasidum* isolated from *Pterostylis grandiflora* on SSE agar

Radial colony growth and dry weight biomass provide useful measures to compare growth responses to varying media and environmental conditions. As fungi are adapted to grow on solid substrates, it is recommended that mechanisms involved in the control and regulation of mycelial growth are better studied on solid medium than in liquid culture (Prosser and Tough 1991). Radial growth rate, defined as $Kr = (R1 - R0)/(t1 - t0)$, where $R0$ and $R1$ are the colony radii at time $t0$ and $t1$, respectively (Trinci 1971), is a ready measure of comparative fungal growth.

Traditional measures of fungal biomass include determination of the hyphal density by counting the number of hyphae intersecting an arc of a known length; or determination of the length and diameter of hyphae and assuming the density and water content. However, these indirect methods are time-consuming, and simple and accurate methods are now available for the determination of the fungal colony biomass dry weight.

Procedure 3.3

Determining orchid mycorrhizal colony biomass **(Reeslev and Kjoller 1995)**

1. In a laminar flow, dissect an active fungal colony from a nutrient agar petri dish and melt the agar gel in 100 to 200 ml of distilled water in a water bath at 60°C for 10 minutes.
2. Collect the colony using a pre-weighed membrane filter (pore size, 1.2 mm). The filter and filter holder should be pre-warmed by adding 50 ml of 60°C water to the filtration unit approximately 0.5 minutes before the filtering and washing gently to dissolve the agar from the hyphae.
3. The filters are then dried at 80°C overnight.
4. Use uninoculated plates as a control to correct for any precipitations from the medium or re-solidification of gel material. Data may need to be corrected for any weight loss attributed to the filters.
5. The biomass density of a colony can be calculated as the dry weight (in milligrams) divided by the area of the colony (in square centimeters). Six replicates should be used.

3.8 MEDIA FOR FUNGAL CULTURE

The preparation of media, from the mixing of components to the setting of the poured media, takes approximately three hours. If 2 L of media is made at the same time, then a hot plate or water bath can be used to keep it warm while poured plates are setting in the laminar flow.

Procedure 3.4

Fungal isolation medium (FIM)

For 2 L:	Sodium nitrate (NaNO3)	0.6 g
	Potassium dihydrogen orthophosphate (KH_2PO_4)	0.4 g
	Magnesium sulphate heptahydrate ($MgSO_4.7H_20$)	0.2 g
	Potassium chloride (KCl)	0.2 g
	Yeast extract	0.2 g
	Sucrose	5.0 g
	Agar	16.0 g

In a microwave-proof jug add all components, except streptomycin sulphate (see following paragraph) and bring up to 2 L with distilled (DI) water. Mix and adjust pH to 6.8 with drops of either 1M HCl or NaOH using a Pasteur pipette. Add agar and dissolve components in microwave for approximately 10 minutes. Pour equal volumes into 500 ml stock bottles and autoclave for 20 minutes at 121°C. Cool enough to touch in a 50°C water bath before adding the streptomycin sulphate (see following paragraph) under sterile conditions. Pour into Petri dishes in a laminar flow cabinet. Allow FIM plates to cool before replacing lids.

Stock streptomycin sulphate solution.

Dissolve 1 g of streptomycin sulphate in 70 ml of deionized water. Store in the dark at 4°C.

Procedure 3.5

Soil solution equivalent (SSE) media with streptomycin sulphate

For 2 L:	MES (2–(N–morpholino)ethanesulfonic acid) buffer	0.4 g
	DI water	1400 ml
	Stock solutions A, B, and C	200 ml each
	Sucrose	4 g
	Agar	16 g

Add the MES buffer to the DI water. Measure out each stock solution and pour separately into the MES buffer/water solution. Add sucrose (as a carbon source) and agar to the solution and adjust the pH to 5.5 by adding drops of 1M HCl or NaOH using a Pasteur pipette. Microwave the medium to dissolve the agar, stirring every 5 minutes, then pour into 500 ml glass bottles, autoclave for 20 minutes at 121°C at 1.05 kg/cm^2 15–20 psi. After autoclaving, place bottles in hot water bath until cool enough to handle. Add 5 ml of sterilized streptomycin sulphate stock solution per 500 ml bottle using a syringe and sterile 0.22 µm millipore filter attachment. Mix well and pour media into Petri dishes under sterile conditions.

Stock Solutions

Prepare three stock solutions by weighing the quantities of salts in the table and placing them into stock bottles (Table 3.4). Add 1 L of distilled water and mix thoroughly. Separate stocks are required to avoid reactions. Stocks will make 10 L of media and should be stored in a refrigerator.

Table 3.4 Stock solutions for SSE media

	Common name	g/L	g/500 mL
Stock A			
NH_4NO_3	Ammonium nitrate	0.4	0.2
KH_2PO_4	Potassium dihydrogen orthorphosphate	0.0136	0.0068
$MgCl.6H_2O$	Magnesium chloride	0.61	0.0068
NaCl	Sodium chloride	0.058	0.029
Stock B			
$CaSO_4.2H_2O$	Calcium sulphate	0.861	0.4305
Stock C			
FeEDTA(Na)	Iron	0.073	0.0365

Procedure 3.6

Potato dextrose agar (PDA)

For 2 L:	PDA powder (1/2 label rate)	13.6 g
	Agar	12 g
	DI water	2000 ml

Add PDA powder and agar to the DI water in a microwave-proof jug. Adjust pH to 6.8 by adding drops of either 1M HCl or NaOH using a Pasteur pipette. Microwave on high until the agar has dissolved and pour into 500 ml stock bottles before autoclaving for 20 minutes at 121°C at 1.05 kg/cm^2 15–20 psi. Place the bottles in a water bath until cool enough to handle, then pour into Petri dishes under sterile conditions. We have found that one-half of the recommended label rate is suitable for maintaining orchid mycorrhizal cultures so that they do not become too vigorous while in storage.

Procedure 3.7

Oatmeal agar (OMA)

Oatmeal agar is prepared for in vitro symbiotic germination trials (Chapter 4) or for the underlying agar source of agar growth containers for symbiotic orchid propagation (Chapter 6). The oats provide a carbon source for the mycorrhizal fungus.

For 2 L:	Crushed rolled oats	5 g
	Agar	16 g
	DI water	2000 ml

Add the rolled oats and agar to the DI water, make up to 2 L and mix well. Adjust pH to 5.5 by adding drops of either 1M HCl or NaOH using a Pasteur pipette. Microwave the solution for 15 minutes to dissolve the agar, stirring every 5 minutes. Pour the media into four 500 ml glass stock bottles and autoclave for 20 minutes at 121°C at 1.05 kg/cm^2 15–20 psi. Place the bottles in a hot water bath until cool enough to handle, then pour into Petri dishes (or agar growth containers) under sterile conditions. Note: breakfast oats without additives are suitable.

Procedure 3.8

Melin-Norkrans liquid media (CN MNM)

This modified low carbon and nitrogen media is typically used for growing cultures for DNA extraction or revival from water storage.

For 1 L	Glucose	5.0 g
	Potassium dihydrogen phosphate (KH_2PO_4)	0.3 g
	Diammonium hydrogen phosphate ($(NH_4)2HPO_4$	0.25 g
	Magnesium sulphate heptahydrate ($MgSO_4.7H_2O$)	0.14 g
	Calcium chloride dihydrate ($CaCl_2.2H_2O$)	66.23 mg
	Sodium chloride (NaCl)	25 mg
	Zinc sulphate (ZnSO4)	3.0 mg
	Ferric ethylenediaminetetraacetic acid (EDTA)	12.5 mg
	Thiamine	0.13 mg

Combine all components except ferric EDTA and thiamine in 1 L of DI water and stir on a shaker at 2.5 rpm to dissolve. Adjust the pH to 5.0–6.0 by adding drops of either 1M HCl or NaOH using a Pasteur pipette before the addition of the ferric EDTA. Pour the media into four 500 ml glass stock bottles and autoclave for 20 minutes at 121°C at 1.05 kg/cm^2 15–20 psi. Once autoclaved and cooled, add thiamine, then pour into 30 ml tubes under sterile conditions.

Procedure 3.9

Pouring media

There are a number of approaches for pouring media into Petri dishes (see Figure 3.17). Below is a general guide as to how this procedure can be done, remembering the important rules for laminar flow use as previously detailed:

1. Wait for media in stock bottle to cool sufficiently to handle comfortably. Agar poured while still very hot will create excessive condensation when Petri dish lids are replaced.
2. Media can be poured into Petri dishes in different ways. Two are described here:
   a. Remove Petri dishes from sealed bags and stack on one side of the laminar flow cabinet. Open Petri dish lid and pour agar, moving left to right (if right-handed) to avoid any part of the hand or clothing hovering over an open jar or dish. Replace lid when set and any condensation has evaporated.
   b. Place Petri dishes in rows, one at a time, beginning from the back of the laminar flow to the front. Prior to pouring, the dishes are positioned in the row with lids three-quarters removed and positioned toward the pourer. Once the first row is poured, the lids are replaced to cover 90% of the Petri dish, allowing any condensation to evaporate before covering completely. Usually three rows will fill the cabinet, depending on its size.
   c. Unless using immediately, stack Petri dishes and place them back into the original bag and seal the top or wrap tightly with cling wrap. Store in a cool, dry container, and use within three months.

Figure 3.17 Orchid conservation volunteers pouring media into Petri dishes in a laminar flow cabinet

Notes:

- Prior to opening the bag containing the Petri dishes, it is a good practice to squeeze the bag tight and if any air is expelled, it is possible the bag may have an air leak and the Petri dishes may not be sterile.

- If there is a contamination risk during the pouring process, it may be worthwhile to leave the poured plates for a week. If contamination is present, discard the plates.
- If the plates have been stored for a considerable period of time prior to use, carefully check each plate for contamination, particularly around the edges, as the plates can sweat and carry contamination through the condensation.

4

ORCHID SEED GERMINATION TECHNIQUES

4.1 BACKGROUND

The seed of almost all wild orchids have an obligate requirement for a mycorrhizal fungus to facilitate germination. Although the dependency of the requirement and the specificity of the fungus involved varies considerably between taxa, mycorrhizal-assisted nutrition is essential given the nature of the dust-like seeds that are characteristic of all orchids. The establishment of self-sustainable orchid populations from artificially propagated seedlings will ultimately require the management and persistence of their mycorrhizal partners to ensure the survival of out-planted seedlings and to enhance further seed-based recruitment. As such, the methods detailed in this chapter deal specifically with symbiotic germination of orchid seeds. It is well established that orchids can germinate asymbiotically (without mycorrhiza) under laboratory and glasshouse conditions; and two case studies are included here showcasing those methods. Many orchids, if transferred from an asymbiotic culture to potting mix or substrate (in the case of epiphytic species), have the ability to assimilate the appropriate mycorrhizal fungus from the mix, including many orchids with mycorrhiza present in their roots.

4.2 MYCORRHIZAL FUNGI FOR SEED GERMINATION OF TERRESTRIAL ORCHIDS

Understanding the dynamics of the underground phase in terrestrial orchids is fundamental for identifying drivers of rarity and making informed decisions for conservation strategies. Orchids produce large numbers of dust-like seeds (Figure 4.1), lacking many of the storage reserves found in other seed-bearing plants (Arditti and Ghani 2000; Batty et al. 2000). Consequently, this and the underground germination of terrestrial species led to the understanding that mycorrhizal fungi are necessary for seed germination. Key studies by Warcup (1973; 1975; 1981) and Ramsay et al. (1986) on Australian terrestrial orchids demonstrated the necessity of a mycorrhizal fungus to initiate germination and seedling development. This has also been shown for a wide range of terrestrial taxa from North America (McCormick et al. 2004; Zelmer et al. 1996), Northern Europe (Rasmussen 1995), and Mediterranean taxa (Jacquemyn et al. 2011).

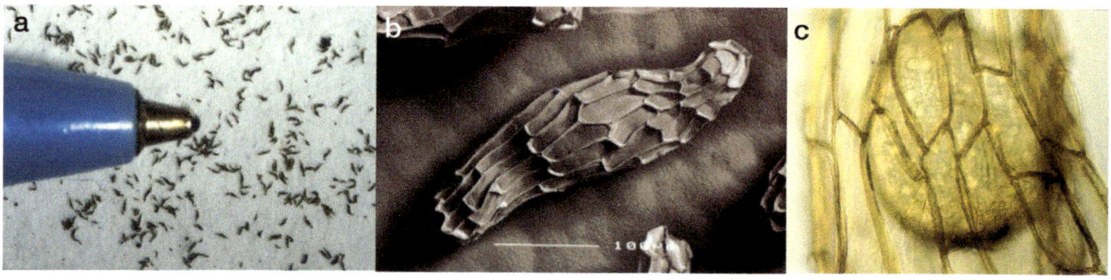

Figure 4.1 Properties and structure of dust-like orchid seed: (a) scale of an orchid seed shown against a pen, (b) Scanning electron microscope image of a seed coat, and (c) structure of the proembryo inside seed coat.

Although the mycorrhizal fungus requirement for seedling germination in terrestrial orchids is well-documented, the degree of dependency in mature plants is less clear. It is likely that orchid-mycorrhizal dependency varies according to the autotrophic (green-leaved orchids) to heterotrophic (nongreen orchids) lifestyle of the orchid (Smith and Read 1997). For green orchids, as the orchid matures, dependency on its mycorrhizal fungus changes to partial mycoheterotrophy with a decreased dependency correlating with an increased photosynthetic capacity of the leaves. Some orchids remain nonphotosynthetic throughout their lives and as a result; life-long mycoheterotrophism is essential for the survival of these species (McCormick et al. 2004). Such species, also known as saprophytic orchids, may associate with a range of specific autotrophic (green) plants, thus engaging in complex and highly specialized tripartite interactions. For example, *Corallorhiza maculata* and *C. mertensiana* associate with fungal species belonging to the ectomycorrhizal family Russulaceae (Taylor and Bruns 1999).

It is likely that a continuum exists in the specialization of the mycorrhizal requirement for germination from highly specific to generalist (Swarts and Dixon 2009) (Figure 4.2). In a study investigating symbiotic germination with a range of terrestrial taxa from different genera, complex webs of compatibility (Bonnardeaux et al. 2007) were found. Two weed-like (disturbance-tolerant invasive) orchids, *Disa bracteata* and the native Australian *Microtis media*, had the broadest network of mycorrhizal fungal partners; both being capable of germinating on a wide range of mycorrhizal types. In contrast, other native orchids were compatible with a narrow range of fungal partners (*Diuris magnifica* and *Thelymitra crinita*), or germinated exclusively with their own fungus (*Caladenia falcata* and *Pterostylis sanguinea*). European terrestrial orchids in the genera *Orchis, Ophrys, Dactylorhiza*, and *Serapias* used a range of fungal partners, with more common species compatible with a broader range of fungi than rare species of the same genus (Muir 1989).

The achlorophyllous Western Australian underground orchid *Rhizanthella gardneri* forms a highly specialized tripartite association with a *Rhizoctonia* mycorrhizal fungus that operates as an ectomycorrhiza on the roots of a nearby myrtaceous shrub, *Melaleuca uncinata* (Warcup 1985). This specialized association may be a key factor in the rarity of the orchid, compounded by extensive land clearance of the orchid habitat for farming.

Some orchids, such as *D. bracteata* and *M. media*, form relationships with a diversity of fungi, with specificity decreasing with time. Some fungi support protocorm formation, but are unable to support subsequent seedling growth (Bonnardeaux et al. 2007); whereas other orchids switch fungal partners during the life of the orchid, as found by Xu and Mu (1990)

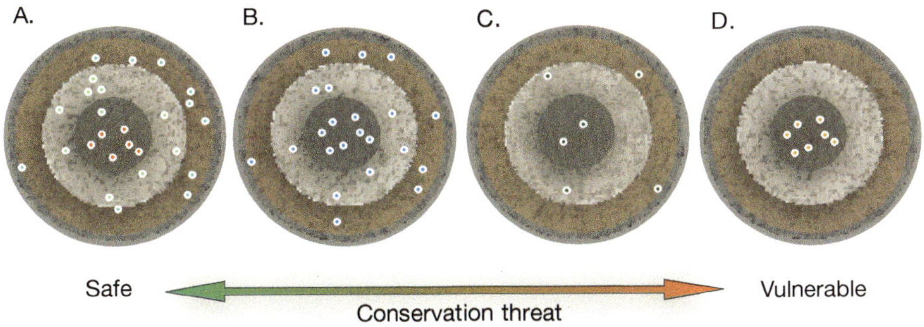

Figure 4.2 Fungal specificity can limit habitat expansion capacity and may dictate whether an orchid species is rare or common. For common species, two mycorrhizal scenarios are possible: (A) by associating with a variety of mycorrhizal partners, orchid occupancy can occur across a variety of habitats and facilitate commonness or, (B) one ecologically robust fungus over multiple habitats may facilitate habitat expansion. In comparison, an orchid may be rare if it has either a highly specialized association with one ecologically diverse but rare fungus, resulting in the orchid occurring across multiple habitats (C) or (D), where an orchid is specialized on one mycorrhizal partner restricted to a few locations in one habitat type, and therefore represents a species at the highest conservation risk in terms of impact of habitat change.

for *Gastrodia elata* where the mycorrhizal fungus changed from *Mycena* to *Armillaria* as the plant matured. In some cases, seed germination has failed when germination is attempted using mycorrhiza isolated from adult plants (Rasmussen 2002), though failure may be due to the isolation of nonsymbiotic fungi.

4.3 PROPERTIES OF ORCHID SEEDS

Of all the angiosperms, the seeds of the Orchidaceae are among the smallest and lightest, and are produced in the largest numbers. Measuring just 0.07 to 0.4 mm across and 0.11 to 1.97 mm in length (Arditti and Ghani 2000), orchid seed has been described as dust-like. The weight of a single orchid seed varies considerably from 0.31 μg to 24 μg depending on the species. These size and weight dimensions allow many seeds to be contained in a capsule, from the smallest (capsules) (20–50 seeds) to the largest (4,000,000 seeds per fruit). *Cattleya*, for example, produce one of the largest seed capsules in the family—each capsule containing many hundreds of thousands of seeds (Arditti and Ghani 2000). Although a single seed capsule produces many seeds, very few germinate and even fewer develop into mature plants. A study of *Caladenia arenicola* seed germination found that each seed capsule contained up to 30,000 seeds (Figure 4.3), however, only a few seedlings were likely to succeed to maturity under natural conditions (Batty et al. 2001a).

For terrestrial orchids, the underground phase of the mycorrhizal association has been difficult to research due to the minute size of orchid seed and the difficulty in tracking seed movements from the time of seed release. Epiphytic orchids are better understood, as protocorms can be observed as small white, often greenish, globoid structures wedged into bark or on lichens/mosses. Through *in vitro* germination experiments (found later in this chapter) and new seed retrieval techniques including direct seed baiting from field sites (Chapter 5), more information is being gathered on the germination biology and

Figure 4.3 *Caladenia* seed capsule (photo: Andrew Bould)

physiology of orchid seeds. In addition, seed dispersal experiments and molecular analysis have greatly contributed to knowledge on the dispersal characteristics and physical properties of these tiny seeds. Production of large numbers of dust-like seeds favors high dispersal rates reflected in the wide genetic variability often found in orchids (Batty et al. 2002; Zettler et al. 2003).

Testas (seed wall) of orchid seeds are usually transparent from smooth to reticulated, with ornamentation varying between species. Since orchid seeds are so light, they can float on air streams for long periods, enabling dispersal over considerable distances. Although wind transportation is the generally accepted mode of dispersal, given their small size, orchid seed has been known to be dispersed by animals and birds—in fur, feathers, or hair, or via mud clinging to parts of their bodies (Arditti and Ghani 2000).

These physical characteristics of orchid seed have both advantages and disadvantages for the orchid biologist. Small seed in large numbers permit easy collection and large-scale experiments. On the other hand, viability testing of such small seed is difficult. Orchid seeds of some taxa such as *Cypripedium* are also often difficult to germinate and may require the presence of a compatible mycorrhiza or a complex asymbiotic media combined with seed dormancy-breaking pretreatments (such as chilling).

Seed recruitment success varies from species to species in the orchids, with some temperate terrestrial species possessing a requirement for stratification or aging in soil to release seed dormancy (Stoutamire 1974). This implies that the seed of terrestrial orchid species, like many other seed plants, requires optimal environmental windows (moisture, temperature, suitable substrate, lack of predation/pathogens) for germination and seedling establishment. In contrast, many tropical orchids germinate readily in the presence of moisture, nutrients, and favorable germination temperatures with seeds of these species generally lacking physiological barriers to germination (Baskin and Baskin 1998; Rasmussen 1995).

4.4 HAND POLLINATION FOR SEED PRODUCTION

Pollinating agents for orchids are generally insects, sometimes vertebrates (e.g., birds), or may be self-pollinating. Orchids often have high levels of outcrossing as a result of both pollinator behavior and high levels of genetic diversity within populations (Hamrick and Godt 1996). Outcrossing facilitates the preservation of genetic variation which may lead to improved seed set. For small populations that may have difficulty attracting pollinators (such as rare species or isolated populations), pollination success may be low. In these cases, hand pollination may be required to ensure adequate fruit set. This will also be necessary for plants contained in *ex situ* collections, as pollinators are likely to be absent or have no access to plants. If collecting from naturally pollinated populations, there are limitations on the proportion of seed that should be collected. For example, sample no more than 10% of total seed production in a population and do not collect when there are very few individuals. For larger populations or rare taxa (with appropriate approvals), seed production can be enhanced through hand pollination; but limit this to alternate years or different plants, as repeated pollination of the same plant may lead to plant decline (Farnsworth et al. 2006).

To preserve the genetic integrity of the seed to be collected, hand pollination by transferring pollen from individuals between populations is most desirable (Figure 4.4). However, this will be dependent on the objectives for seed germination and storage requirements. Hand pollination can be carried out with fine forceps or a toothpick to remove the pollinia from beneath the anther cap and transferring it to the sticky surface of the stigma of

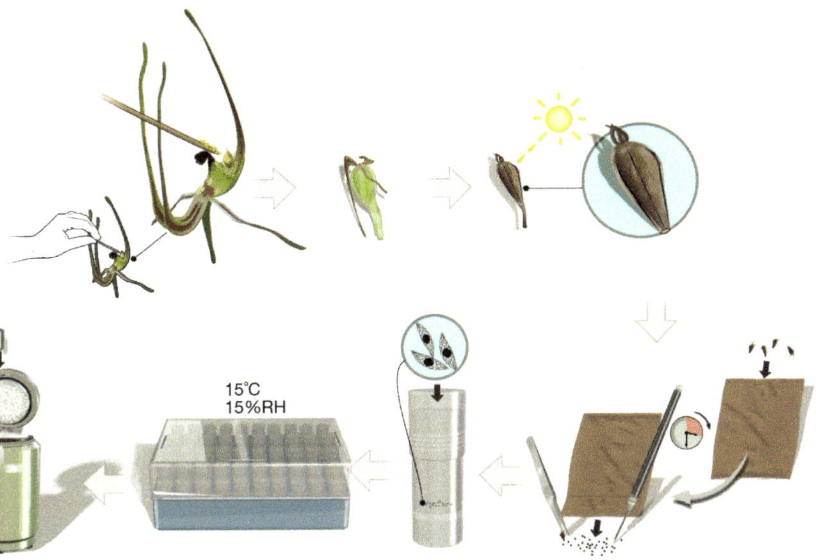

Figure 4.4 Seed collection flow chart showing hand pollination of orchid flowers, the natural maturation of seed capsules, placement of seed in paper envelopes, moving dried seed into small tubes for drying, and finally, placement of seed in low temperature environments such as liquid nitrogen for long-term storage.

another plant. Orchids vary in flower morphology so care must be taken to ensure that the procedure is completed correctly. In some cases the flower may need to be carefully dissected to access the anther cap or the stigma, for example *Cypripedium, Paphiopedilum*, and *Phragmipedium* species. Where pollen quality may be questionable, the "hanging drop" techniques described by Light and MacConaill (1998) may be used to assess pollen viability.

4.5 SEED COLLECTION

Guidelines for seed collection are described in Procedure 4.1 and include information on seed maturity, packaging, and storage (see Figure 4.4).

Procedure 4.1

Collection of mature orchid seed

1. Native orchids are protected species under national and local government legislation for most countries; and include taxa listed under the Convention on International Trade in Endangered Species (CITES), Appendix 1. Before collection, ensure the correct licenses/permits have been obtained and permission has been granted from landowners and permitting authorities.
2. No more than 10% of the natural seed set from wild populations should be collected, and no seed should be taken where there are only a few individuals in the population—unless part of a rare species conservation program. Hand pollination of rare species where natural seed set is limited will generally result in more capsules than would naturally occur in the wild and is a useful strategy to obtain reliable quantities of seed.
3. If hand pollination is required, obtain a good knowledge of the flower morphology of the species to be pollinated.
4. Cross-pollinate individuals between populations at a site to ensure greatest outcrossing potential.
5. Adequately note the position of pollinated plants to facilitate relocation when it's time to collect seed. This can be done with GPS, flagging tape, or a metal peg as described in Chapter 2. It should be noted that this may bring undue attention to the orchids at the site, and particular care should be taken when working with threatened species.
6. Orchid seed pods are suitable to collect when they commence yellowing just prior to the splitting of the capsule. The seed at this stage will be fully developed. Seed capsule maturity varies in time, depending on species and weather conditions; from 4–6 weeks (many temperate terrestrial taxa) to 11 months for epiphytic species. It is recommended to check progress regularly, as capsules will ripen faster following dry and/or sunny periods.
7. Capsules in an inflorescence will vary in maturity depending on flowering time and the pollination event, so it is advisable to collect individual capsules when mature from a number of plants, leaving the remaining capsules to mature. A mesh bag positioned to capture the seeds when the capsule splits can be installed at the site.
8. When capsules are yellowing, remove from the stem and place in a paper envelope. Avoid using airtight or plastic containers, as these will retain moisture and could lead to the development of mold.

9. Freshly collected seed should not be stored in the refrigerator until the seed is dried as high seed moisture will limit seed viability.
10. If capsules are still green, remove the whole inflorescence and place in water in a well-lit, dry position until the capsule shows signs of splitting.
11. Seed should be dried in a seed-drying facility maintained at 15°C and 15% relative humidity (RH). After 10 days in these conditions, the seed is ready for sowing or long-term storage in a seed bank (Chapter 10).
12. Seed label information should include: donor plant details, provenance, harvest date, condition of seeds, GPS location, and herbarium voucher specimen number.

4.6 SOWING MATURE SEED

Where possible, *mature* seed should be collected and sown (as described in Procedure 4.1). The main advantage for this is that mature seed can be stored and banked in an *ex situ* seed collection for later use. Currently, it is not possible to store immature seed although research into this method is continuing. Inspecting seed under a microscope will indicate if the seed is viable through the presence of a pro-embryo (spherical body in the center of the seed). Tetrazolium staining (discussed more in the upcoming text) can establish orchid seed viability, noting that the method may require calibration by germinating the same sample.

Orchid seed can be sown directly onto the agar medium or onto sterile filter paper or other supporting material and transferred to the agar medium. Orchid seeds often have hydrophobic testas and air is usually trapped in the seed, therefore seeds tend to float on the surface of sterilant solutions. To overcome this, a variety of direct sterilizing techniques have been developed, which can include the use of surfactants to release surface tension; vacuum infiltration; and the use of agitators, shakers, and stirrers to suspend seed in the sterilant solution. Sterilized seed must then be removed from the container and test tubes. Pipettes, glass filters, and filtration systems have been designed for this purpose as explained in detail below.

The transfer of sterilized seeds to a Petri dish or germination flask is completed using a sterile technique in a laminar flow (see Chapter 3), though it is possible to sow seed in semi-sterile conditions with care. While epiphytic tropical taxa have received considerable attention, the development of expertise in temperate terrestrial orchids is increasing for both conservation and horticultural purposes. Many stages of seed sowing, transfer, and media preparation can take place in a kitchen or a reasonably equipped laboratory. Hicks (2000) outlines the use of glove boxes and working over steam as cost-effective "home" alternatives to a laminar flow cabinet.

Techniques for sowing epiphytic and terrestrial orchid seed are generally similar, however, terrestrial seeds may require additional sterilization time or different pre-sowing treatments (e.g., acid or chilling) in order to degrade the seed coat and break seed dormancy.

Case Study 4.1

Interface between propagation techniques and conservation studies: germination requirements in Cypripedium calceolus *and* Epipactis palustris

by Hanne N. Rasmussen
University of Copenhagen, Department of Geosciences and Nature Resource Management Forest, Nature, and Biomass

Danish native orchid propagation work began in 1986, inspired by similar activities on native orchids at the Royal Botanic Gardens, Kew, but it gradually evolved from horticultural to conservation research. The first major study showed that through the use of a temperature gradient plate, it was possible to modify the interaction between seed and fungus in symbiotic cultures, and thus the results obtained could be interpreted phenologically. Research then shifted to investigate cold stratification of seeds and to manipulate light regimes, which also affected germination and symbiotic growth that provided important ecological contexts for ongoing research (Rasmussen and Rasmussen 1991). Subsequent introduction of seed packet techniques (Rasmussen and Whigham 1993) has made observation of these processes more direct, and supplements the *in vitro* approach. Conditions *in vitro* differ in several respects from those in nature, and caution is required when drawing conclusions:

- Seeds need to be surface sterilized, and the seed coat is thus chemically changed compared with natural seeds. Using immature seeds in order to avoid seed coat inhibition is also artifactual.
- The sterile environment eliminates all other organisms other than the mycobiont.
- The microclimate is stable and humidity high within the culture vessel compared with the soil environment.
- The substrate is usually an agar with a composition dissimilar to natural soil. To provide a substrate more akin to soil simplification of substrates (*minimal media*) can be used to approach natural conditions (Figure 4.5a).

Polymorphism in orchid seeds is a largely neglected phenomenon with ecological implications: varying seed germination requirements within a seed batch provides ecological *bet-hedging*, which increases the probably of an establishment event. It may be difficult to identify overlapping or separate guilds within a seed batch, and most published research is based on the successful germination of just a small fraction of seeds. In our asymbiotic testing of *Cypripedium calceolus* L. *in vitro*, one cold treatment resulted in sporadic germination (2.2%). After 12 months, all seedlings were removed from the dishes and the remaining seeds subjected to a second cold treatment where 29% germinated (Pedersen et al. 2012; Rasmussen and Pedersen 2011). Thus, seeds reacted differentially to chilling, which might represent ecological requirements for either one or two winters. Yet 70% of seeds remained dormant indicating the potential for other ecological cues to break dormancy.

Where there is difficulty isolating the mycobiont, asymbiotic culture, as mentioned previously for *C. calceolus*, may be the only option. However, asymbiotic germination may greatly underestimate the effect of climatic factors. Thus, in germination of *Epipactis palustris* (L.) Cr. (Figure 4.5b, c) we found highly significant interaction between cold stratification and mycorrhizal presence (Rasmussen 1992), with germination rising above 50% when the treatments

Figure 4.5 (a) Seeds of *Goodyera pubescens* sown on water agar (8 g of agar in 1 L of water), to which a scoop of sterilized wood powder was added. Hyphae from the inoculum extend to the seedlings. Both seedlings and mycobiont developed on wood debris as carbon source. The culture was deprived of light. (b) *Epipactis palustris* whose seeds required both a cold period and signals from a mycobiont in order to germinate *in vitro*. This suggests that germination in nature takes place in spring. (c) *Epipactis palustris* in its natural habitat where there is normally year-round moisture (photos: Hanne Rasmussen).

were combined. Separately, both chilling and mycorrhiza had only a limited effect (<5%), whereas in combination, the effect was ecologically sound. Overall, the Danish orchid conservation program has progressed, built upon a sound foundation of a combination of *in vitro* and *in situ* science.

4.6.1 Seed Viability Staining

The microspermous nature of orchid seed and the difficulties in germinating many species due to their mycorrhizal requirements and potential dormancy issues make seed viability assessment particularly challenging for orchids, in comparison to other seed-bearing plants where chemical staining techniques can be calibrated via direct germination measure.

With this caveat in mind, viability of orchid seed can be tested by seed germination counts (SGC) on asymbiotic media or by chemical staining to determine if ungerminated seed is inviable or has retained dormancy. In comparison to direct germination assays, staining techniques have the advantage of being rapid, but they do need to be calibrated with seed germination.

To be effective, the stain must be brought into direct contact with the pro-embryonic cells (the globoid body within the seed responsible for germinating into a protocorm). In some taxa, the seed must first be made permeable to water, or, alternatively, the embryo must be released from the testa to facilitate stain uptake. For example, this can be done by presoaking or sterilizing seeds in calcium hypochlorite solution or by squeezing and agitating soaked seeds in a drop of water between microscope slides to expel embryos from the testa (Pritchard 1985).

The most common chemical stains to determine viability of orchid seeds are triphenyltetrazolium chloride (TTC) (see Procedure 4.2) and fluorescein diacetate (FDA) (Lakon 1949; Pritchard 1985). Viability tests using these stains have been successful with epiphytic tropical orchids (Singh 1981), and Van Waes and Debergh (1986) adapted such an assay for European terrestrial orchids. The number of seeds that incorporate stain should correspond to the number of germinable seeds (the latter must be lower than the former for the stain test to be considered accurate). For example, Batty et al. (2001b) used TTC, FDA, and Evans blue in a comparative study with symbiotic germination tests; finding no clear relationship between seed scored as viable and the percentage of symbiotic germination for each of the stains tested. In another example, Lauzer et al. (1994) found that for *Cypripedium* seed, germination of mature seed was significantly lower than the percentage of embryos with positive staining patterns with TTC. These studies highlight the need for further histochemical research into the physiological processes that control dormancy and seed viability in orchid seed.

Procedure 4.2

Viability testing of orchid seed using TTC

Preparation of calcium hypochlorite solution (5% weight per volume solution)

1. Add 5 g calcium hypochlorite powder to 100 ml of water; mix together using a mortar and pestle as $Ca(OCl)_2$ is difficult to dissolve.
2. Stir in the remainder of the water and 1 ml Tween-80 (1% v/v).
3. Stir for 10 minutes, then filter. Use filtered solution immediately.

Preparation of the TTC solution

1. Dissolve 1 g of TTC in 100 ml distilled water.
2. Adjust pH to 7 with 0.1 M NaOH.
3. Store in the dark at 20°C (foil wrapped brown glass bottle at room temperature). TTC is toxic and must be handled with gloves and prepared in a fume hood.

Viability testing

1. Place a small amount of seed in 10 ml of $Ca(OCl)_2$ solution in a pyrex test tube with a screw cap. Gently shake for 3 to 20 minutes depending on genus. Note that for some species, up to 3–4 hours may be required.
2. Seeds will usually sink to the bottom of the tube.
3. Pipette off bleach solution and replace with sterile water. Change water at least twice and then leave seeds in water for a minimum of 24 hours.
4. Pipette off water and replace with 10 ml of 1% TTC solution. Place tubes in darkness at 30°C for 24 hours.
5. Decant TTC solution and wash seeds 3 times for 5 minutes each in sterile distilled water.
6. Decant water, then place seeds on filter paper discs divided into quarters.
7. Rinse remaining seeds with a small amount of water and disperse seeds evenly over the surface.
8. Using microscope, score: total number of seeds, number of seeds with pro-embryos, number with pro-embryos stained red.
9. Determine the mean and standard deviation based on four replicates.
10. Marking a grid on the filter paper and using a tally counter will aid scoring. The staining result indicates the maximum possible viability of that seed sample.

Case Study 4.2

Ex situ conservation and reintroduction of Hong Kong's threatened orchids

Stephan Gale, Kelvin Yeung, and Chow Pak Tai
Kadoorie Farm and Botanic Garden, New Territories, Hong Kong

The Chinese territory of Hong Kong lies at the eastern margin of the Indo-Burma Biodiversity Hotspot in the seasonal Asian tropics and has a native orchid flora of around 130 species in 60 genera. Ten threatened species of global conservation importance have been prioritized for *ex situ* propagation and reintroduction by the local environmental NGO Kadoorie Farm and Botanic Garden; eight of these are terrestrial (*Acanthophippium gougahense*, *Calanthe masuca* var. *sinensis*, *Cheirostylis monteiroi*, *Goodyera seikoomontana*, *Habanaria leptoloba* (Figure 4.6a), *Hetaeria youngsayei*, *Paphiopedilum purpuratum*, and *Tainia ruybarrettoi*); and two are lithophytic or epiphytic (*Bulbophyllum bicolor* (Figure 4.6b) and *B. kwangtungense*). Most of these species have been reduced to just a handful of plants throughout their former range as a result of habitat destruction and poaching.

Open pollinated capsules are collected from wild populations or, where necessary, from hand-pollinated plants of known provenance in cultivation (Figure 4.6c). Experimental pollinations have revealed self-incompatibility and inbreeding to be issues affecting seed set and seed quality for several species, notably *A. gougahense*, *B. bicolor*, *B. kwangtungense*, and *P. purpuratum*, with both factors being further exacerbated by pollinator- and mate-limitation in the wild.

Figure 4.6 (a) *Habenaria leptoloba*; (b) *Paphiopedilum purpuratum*; (c) micro-propagation at Kadoorie Farm and Botanic Garden; (d) pollination studies of *Bulbophyllum bicolor*; and (e) reintroduction of *Bulbophyllum bicolor* (photos: Stephan Gale).

Propagation involves collection of immature capsules with the capsule surface-sterilized using a dilute bleach solution, and viability is tested using TTC staining. Micropropagation trials (Figure 4.6d) for all ten species are then done asymbiotically, using three agar-based media (Phytomax P1056—a proprietary product—and two customized Murashige and Skoog blends). Satisfactory germination has been achieved for half of the species (*B. bicolor*, *Calanthe masuca* var. *sinensis*, *Goodyera seikoomontana*, *Habanaria leptoloba*, and *Hetaeria youngsayei*), with P1056 providing the best results. Monitoring of post-protocorm growth and subculturing to fresh media, as appropriate, has helped reduce high mortality associated with certain developmental stages in some species. Deflasking is undertaken from one year (for fast growing species, such as *B. bicolor*), to over two years (for *P. purpuratum*) after germination, with seedlings being transferred to a potting medium of bark chip, coarse perlite, and sphagnum moss, or onto cork in the case of the two epiphytes.

Reintroduction trials at wild sites are most advanced for *B. bicolor*, for which survivorship is approximately 80% after six months (Figure 4.6e). This work is crucial to bolster genetic diversity at the site level, and is being conducted in parallel with investigations of pollination biology, population genetic structuring, and habitat requirements to help inform site selection and planting arrangements.

4.6.2 Sterilants

Commonly used sterilants are sodium hypochlorite (NaOCl) and calcium hypochlorite ($Ca(OCl)_2$) (Figure 4.7) (also see Case Study 4.3 on the semi-sterile technique). Hypochlorite solutions sterilize orchid seed as well as aid in the removal of suberin in the integuments, which enhances water infiltration into the pro-embryo. The suberin content of seed varies between species, and darker integuments usually require longer sterilization times or stronger hypochlorite solutions. While permeability of the testa will increase with progressive oxidation, there is a risk of toxicity under prolonged exposure times which may result in injury or death of the embryo. For new seed collections of untested species, a cautionary approach should be applied beginning first with shorter sterilization periods.

For ease of use and availability, commercially available sodium hypochlorite is usually the sterilant of choice, though a range of sterilants can be used. As the chlorine content of different manufacturer's products are likely to vary, it is important to check the available chlorine on the container. When the available chlorine is 13% as commonly seen, dilutions to achieve a 1% and 5% final concentration of chlorine are made for orchid seed sterilizing. When using household bleach, which typically contains about 5% available chlorine, less dilution is required. Note: avoid use of commercial preparations that contain detergents—check the label for components. Some detergents can damage the pro-embryo.

For species requiring longer sterilization times (up to 24 hours), $Ca(OCL)_2$ is more commonly used than NaOCl, as less embryo damage has been observed for the same duration. To prepare a standard solution of saturated $Ca(OCL)_2$, 10 g powder (24% available chlorine) is dissolved in 140 ml of water. This gives a final solution with available chlorine of 1.7%. $Ca(OCl)_2$ is available in many concentrations and can be found with up to 75% available chlorine. Check the container carefully for the available chlorine content and calculate the amount required.

An alternative treatment to long exposure to $Ca(OCl)_2$ is the use of 0.5–2% sulphuric acid (H_2SO_4) for 5–15 minutes as a pretreatment before sterilizing in hypochlorite (Malmgren

Figure 4.7 Hypochlorite solutions for sterilizing orchid seed

1993). The concentration and duration of acid pretreatment needs to be finely adjusted for each species, as does the time in hypochlorite, with acid-treated seed requiring less sterilizing time.

Case Study 4.3

Semi-sterile techniques for orchid germination

Betty Mauliya Bustam
Syiah Kuala University, Mathematics and Natural Sciences Faculty

Here I outline a semi-sterile technique for germinating orchid seeds that will enable a more user-friendly and practical method for sowing and germinating orchid seeds—at home and for beginners. The technique requires no expensive laboratory equipment, such as Laminar Air Flow (LAF) units or autoclaves, and the approach is safe and cost-effective.

The principle of the semi-sterile technique is based on the use of a widely available antimicrobial sterilant known as Milton, which has been found to be most effective. Milton is a common product used for sterilizing baby bottles and eating utensils—so very safe with

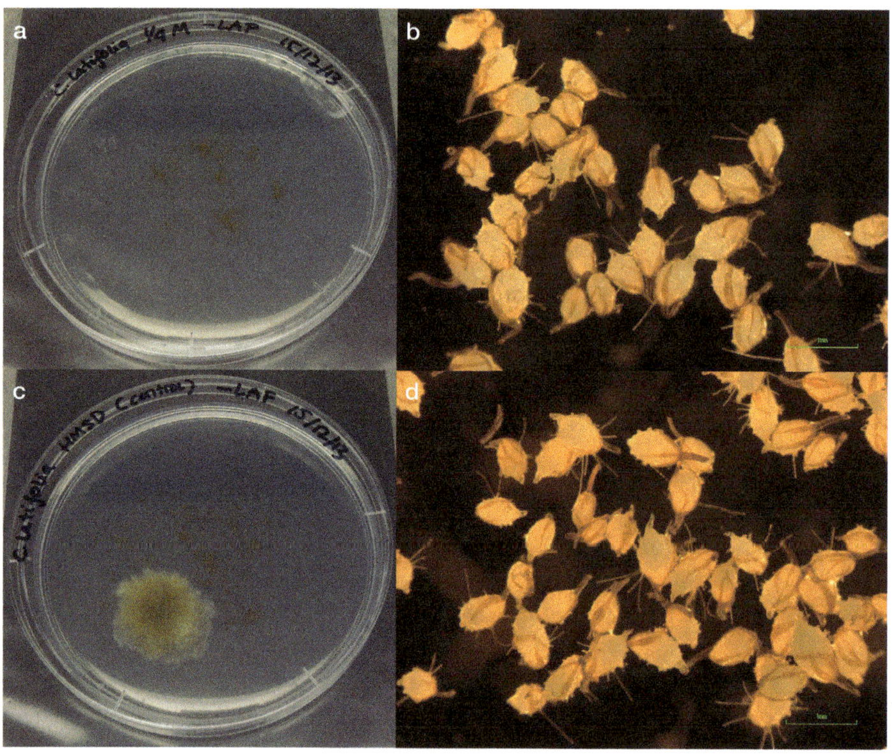

Figure 4.8 Germination plates established using a semi-sterile technique (a) 1 week after sowing with Milton; (b) 6 weeks after sowing with Milton; (c) 1 week after sowing without Milton (control) and; (d) 6 weeks after sowing (control) without Milton but sowing in sterile conditions (control).

the active ingredient based on hypochlorite. The method involves adding the tablet to orchid propagation medium (prepared mixes can be readily purchased online for your particular orchid genus). An effective asymbiotic medium we found useful for terrestrial species was half-strength Murashige-Skoog (MS) basal medium without additional organic compounds or hormones. The medium is abbreviated as HMSD (Bustam et al. 2014) to which 20 g L^{-1} sucrose and 8 g L^{-1} agar were added. All ingredients are then heated with regular stirring on a stove or microwave to dissolve the sucrose and agar. A quarter of a standard Milton tablet is added to 1 L of medium cooled to 50°C. Ensure the tablet is completely dissolved before pouring the medium into the sowing vessel—clean glass jars with wide mouths are very useful. Though the entire procedure was undertaken outside the laminar flow, there was no contamination found in media utilizing Milton compared to controls without Milton which were all contaminated. Although seeds germinated at higher strength Milton (half, one, or more tablets), the best germination and growth was on HMSD with an additional quarter of a tablet of Milton (Figure 4.8).

This simple and highly effective means of germinating and growing orchids is particularly effective for those orchids with a less critical requirement of mycorrhizal fungi for germination—root colonized terrestrial species and epiphytic species. The technique is also applicable to schools and for community groups and community nurseries wanting to germinate orchids.

4.6.3 Stages of Seed Germination

The process of orchid seed germination *in situ* requires that a seed arrives in close proximity to a compatible fungus. Under *in vitro* conditions, this situation is controlled, but the colonization of the seed under both sets of conditions remains the same. Fungal hyphae enter the seed through the microporous gaps between the integument plates. Once inside the seed coat, the hyphae grow through the suspensor region of the embryo and penetrate the walls of the adjacent cortical cells where they form coil-like structures (pelotons). The presence of pelotons marks the compatibility between the orchid and the invading fungus. A protocorm is formed when the meristematic cells within the seed divide and grow. As the protocorm develops, new cells between the seed epidermis and the existing cortex are invaded that eventually grow, forming the outer cortex and becoming the primary point of infection in the developing protocorm. Rhizoids or trichomes, recognized as outward protrusions from the protocorm, become the interface of hyphal activity between the germinated seedling and the culture medium. At this stage, the protocorm is now almost double its original size with roots and leaves appearing.

The stages of seed germination represented in Figure 4.9 are used for scoring progression in germination. Using a dissecting microscope, seeds are assigned to a development stage, and the totals of each stage are summed to obtain a germination percentage. Germination (i.e., irreversible growth of the protocorm) corresponds to Stage 3 with symbiotically germinated seedlings possessing pelotons in the protocorm. This does not guarantee protocorm development onto the green leaf stage, since a higher degree of host-fungus compatibility may be required for some taxa. The decisive criterion of host-fungus compatibility is the presence of seedlings with leaves, not merely fissuring of the testa (Peterson et al. 1998).

Studies comparing the effectiveness of symbiotic and asymbiotic germination have usually shown that symbiotic germination is more rapid and with higher germination outcomes

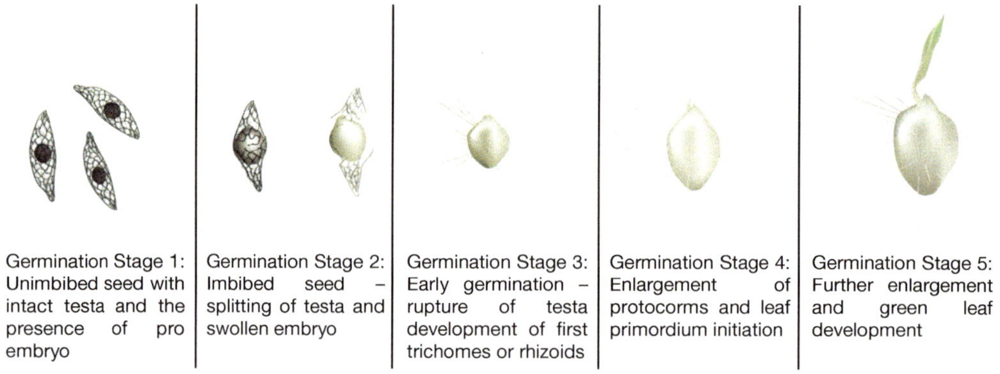

Figure 4.9 Stages of seed germination from seed to germinated seedling, with Stage 5 typically having the presence of a green leaf indicating the orchid is commencing a level of autotrophy (photosynthesis)

than asymbiotic germination (Peterson et al. 1998). In some terrestrial orchids, successful germination can only be achieved by symbiotic germination (Hadley 1982). Smreciu and Currah (1989) showed that *Platanthera hyperborea* seeds germinated and protocorms developed on an asymbiotic culture medium as well as on Warcup's medium with several mycorrhizal fungi. As discussed above, this section focuses on terrestrial orchids, as there is strong evidence for the vital role of orchid fungi in the propagation and growth of most terrestrial orchids. This contrasts with epiphytic orchids, where mycorrhizal fungi may have less importance for seedling growth and therefore, epiphytes are typically propagated by asymbiotic means.

4.6.4 Seed Sowing—Packet Method

The packet method (Figure 4.10), using either filter paper or nylon mesh bags, provides indirect contact with the sterilant and is easier to manage than other techniques. In this technique, only small amounts of seed are required. When using filter paper, higher concentrations of sterilant may be required along with surfactant in order to penetrate the paper. Watch point: the filter paper may disintegrate with longer soaks. The sterilization process remains the same for both types of packets, but the sowing technique varies slightly. For the filter paper method, a high grade of filter paper that will withstand soaking is necessary (e.g., Whatman no. 1, Cat. no. 1001070). For the nylon mesh method, it is important to use a grade that will ensure the seeds will not be lost through the gaps in the mesh (e.g., 90 μm). Seed sowing can take place either on filter paper or in nylon mesh bags, depending on the purpose, before transfer to the nutrient agar medium. Seed sown on filter papers can be used for both large-scale germination purposes or sections of filter paper can be cut and placed on germination plates for simultaneous cross-specificity studies. For example, up to six different orchid taxa can be sown on a plate containing a mycorrhizal fungus.

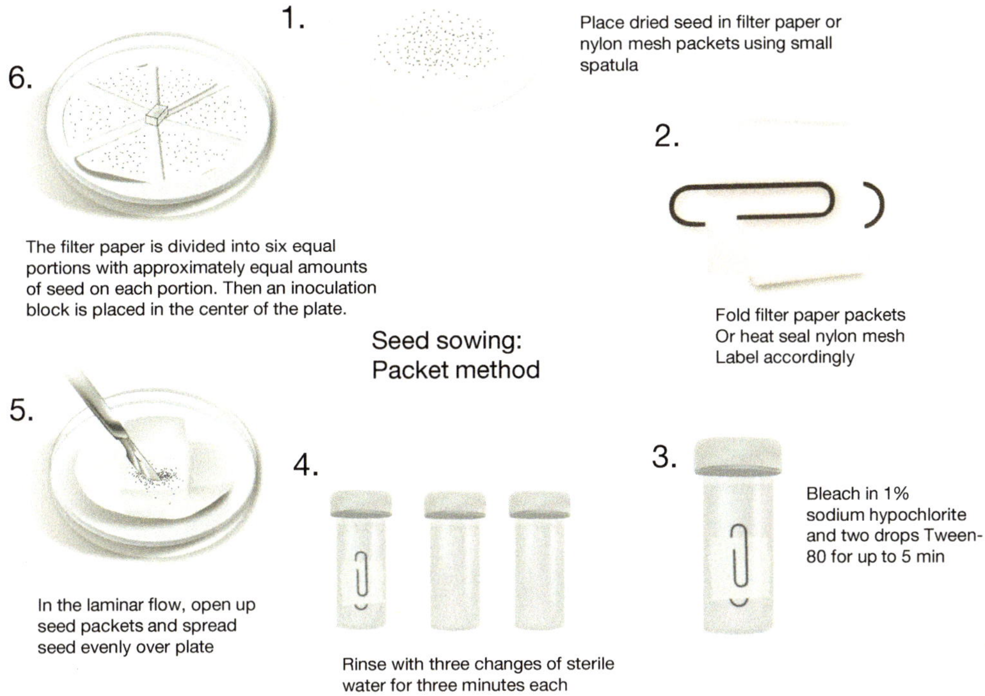

Figure 4.10 Sowing of orchid seed using filter paper packets

Procedure 4.3

***Seed sowing using filter paper packets or nylon mesh* (Figure 4.10)**

1. Prepare seed packets. Fold a piece of 90 mm filter paper (Whatman no. 1, Cat. no. 1001070) into a pocket. Cut nylon mesh and heat seal into 25 mm squares. Using a small spatula, place dried orchid seed in either the filter paper or the mesh. Limit the number of seeds in a packet as the sterilant may not be able to penetrate—make several packets instead.
2. Fold the filter paper to close the pocket and seal with a paper clip. To close the nylon mesh, heat seal the final opening.
3. With forceps, place (a maximum of 5) packets in a jar containing distilled water and remove any air from the packet.
4. Transfer packets to the sterilizing solution (1% sodium hypochlorite) in a beaker or sealable container. Consider adding a few drops of a surfactant such as Tween-80 to ensure the seeds are wet to allow penetration of the sterilant.
5. Place the beaker on a magnetic stirrer and allow to agitate for 2–25 minutes according to the treatment required and the concentration of sterilant.

6. In the laminar flow, remove packets from the sterilant using forceps and place in sterile distilled water (in separate tubes or screw top bottles). Agitate for 5 minutes, repeating this rinse three times.
7. For filter paper seed sowing:
   a. Remove paper clip, and in an empty 90 mm Petri dish, open filter paper pocket with forceps.
   b. Using sterile water and a pipette (or a scalpel), carefully spread the seeds evenly over the filter-paper surface. Here, the bleached seeds may be difficult to see on the white filter paper, so this step may need to be done under a microscope.
   c. With a scalpel and forceps, cut the filter paper into six equal-size portions and place onto Petri dishes containing germination medium (oatmeal agar (OMA), see Chapter 3). Alternatively, lightly dab the filter paper against the medium, holding the filter paper seed-side down. With practice, seeds can be sown very evenly and thinly using this *smearing* process.
8. For nylon mesh bags:
   a. In an empty Petri dish, cut the heat-sealed sections of the nylon mesh and open at the fold.
   b. Dab mesh on the petri dish surface to remove as much liquid as possible.
   c. Lightly dab the mesh downward against the medium, transferring the seeds in a thin layer over the plate.
9. Add the mycorrhizal fungus isolate to the germination plate and leave open in the laminar flow cabinet to dry excess water on the plate surface.
10. Label, close, and seal plates before placing them in the dark at approximately 15°C for four weeks.
11. Check plates on a weekly basis to remove contamination (if any) and presence of outgrowth of the mycorrhizal fungus.

4.6.5 Seed Sowing—Syringe Method

Here the seed comes into direct contact with the sterilant. As all sterilization and rinsing of seed takes place in the same syringe, this procedure can be completed on an open bench, only requiring a clean environment (cabinet or working over steam) for the final sowing stage.

Procedure 4.4

***Seed sowing: syringe method* (Figure 4.11)**

1. In a laminar flow, press a small piece of cotton wool in the bottom of a 10–20 ml glass syringe with a point at the tip. The compacted cotton wool acts as a filter during sterilizing and rinsing stages, but must be loose enough to dislodge for later seed sowing.
2. Add dried orchid seed to the syringe with a spatula.
3. Replace plunger and draw up sterilizing solution (1–2% sodium hypochlorite).

4. Tap the syringe to remove air bubbles and disperse seeds; invert the syringe and depress plunger sufficiently to dispel all air.
5. Agitate by tipping the syringe backward and forward for the minimum of 10 minutes.
6. Force out the sterilizing solution, pushing the plunger to the bottom of the syringe.
7. Pull the plunger to the top of the syringe and rinse the plunger with sterile water.
8. To rinse seeds, draw up sterile water, agitate briefly and force out water using the plunger. Repeat rinsing process 2 or 3 times.
9. Dislodge the cotton wool plug by tapping or pushing the sterile needle up the syringe tip. Then, draw up the small amount (1–2 ml) of sterile water.
10. Disperse seeds in the solution, then squirt seeds onto the medium. Withdraw excess water into the syringe. Alternatively, seeds may be sown by removing cotton wool with forceps and dabbing on the surface of the medium, or by dabbing the end of the plunger.

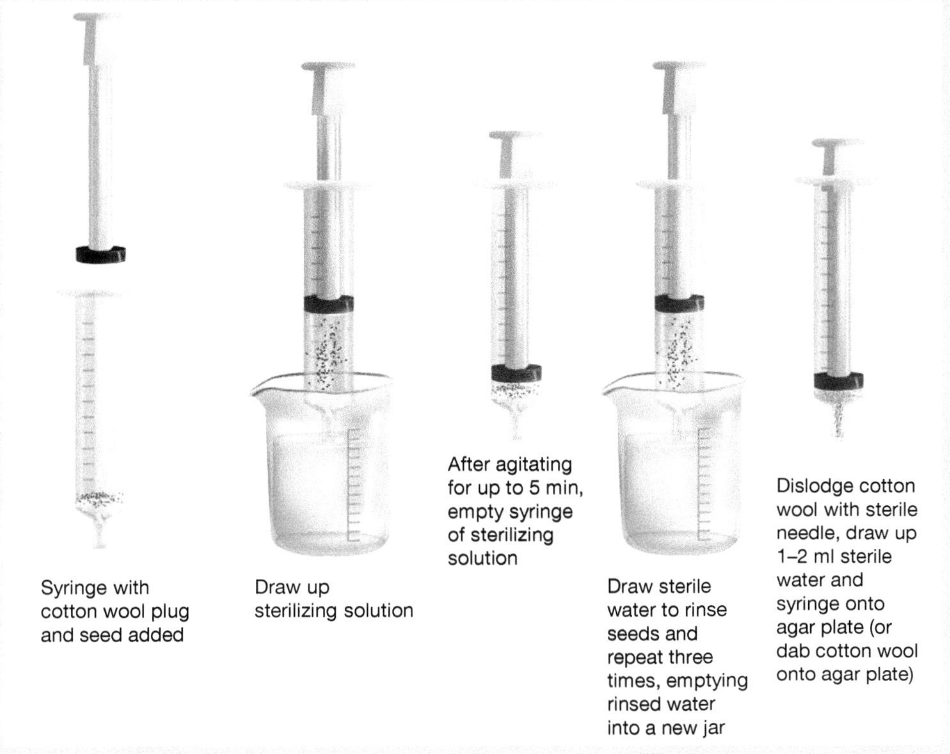

Figure 4.11 Seed sowing: syringe method

4.6.6 Seed Sowing—Direct Sterilization

The test tube method can be used in combination with the tetrazolium viability staining technique (previously mentioned) to obtain a direct comparison between viability testing and germination rates. This is possible, as seed has undergone the same sterilization treatment for both methods. As an alternative to removing seeds from the sterilant, a Buchner funnel with filter paper can be used (Hicks 2000; Thompson 1996). Seed collected on the filter paper can be sown by dabbing the filter paper, seed-side down, on the surface of the medium.

Procedure 4.5

Direct sterilization in test tube **(Figure 4.12)**

1. Using a spatula, place a small amount of seed in a test tube that can be sealed with either a screw-on lid or a rubber bung.
2. Add sterilant with wetting agent to aid contact with the seed and to limit flotation and then seal closed.
3. Shake tube for 2–15 minutes (depending on species) so seeds have good contact with the sterilant. For species that require longer periods of sterilization, such as dark terrestrial orchid seeds, equipment such as a blood rotator will allow constant gentle agitation (for multiple batches an orbital shaker can be used for agitation).
4. Stop agitation 1–2 minutes before sterilization period is completed to allow seeds to settle. Seeds will either fall to the bottom or form a ring around the top of the test tube.

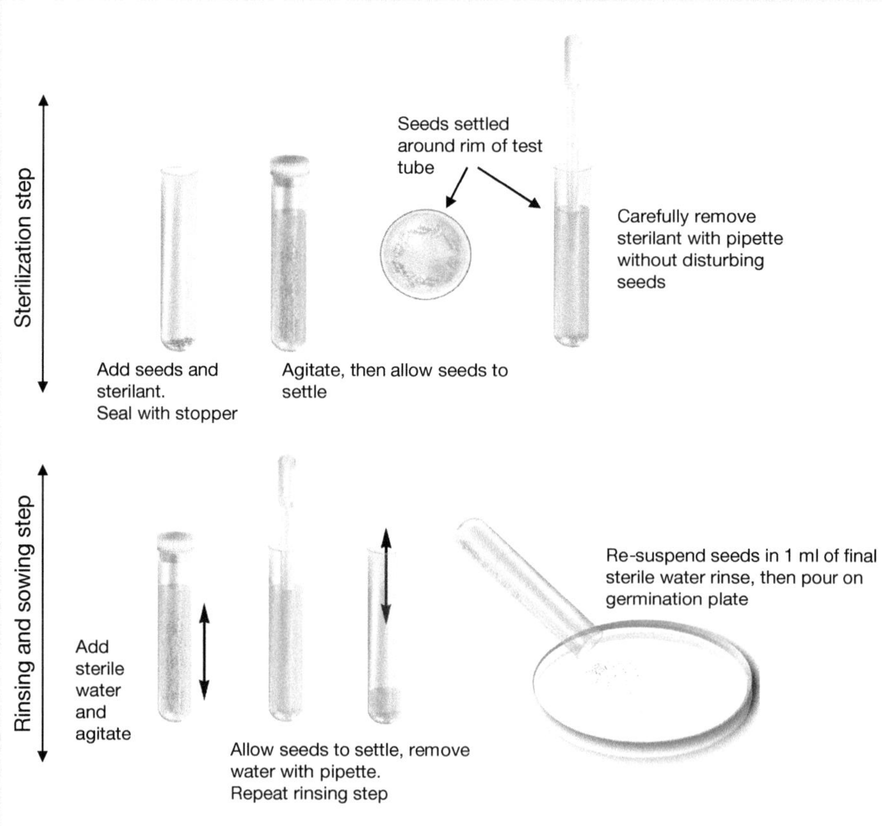

Figure 4.12 Direct sterilization of orchid seed for seed sowing experiments

5. With a pipette (sterile disposable plastic Pasteur pipettes are useful), remove sterilizing solution from above the seed. It is not critical to remove all the sterilant, as any remaining will be progressively diluted through the rinsing steps.
6. To rinse the sterilized seeds, add sterile distilled water, seal, and agitate. Once the seed has settled, remove the water with a pipette. Complete this rinsing step three times, and in the final rinse, leave approximately 1 ml of water in the tube.
7. In a laminar flow, suspend the seeds with a quick shake, then tip test-tube contents onto a Petri dish/flask or use pipette to suck up the suspension and dispense.
8. Spread the sterilized seed over a plate containing OMA media with a spatula or scalpel blade, removing excess water with the pipette.

4.7 SOWING TECHNIQUES FOR IMMATURE SEED

Sowing immature seed is a strategy that can be used if there are problems with obtaining germination of mature seed because of possible seed-germination inhibitors or dormancy issues. This is achieved through the collection of *green* (immature) capsules from the orchid plant before the seed is fully ripe. In general, green capsules are relatively straightforward to sow because the contents of the capsule are sterile. Some disadvantages of using immature seed capsules include the need to sow all the seed in one session and the inability to store collected seed, as well as an increased risk of virus carryover if maternal tissues are in contact with the dissection equipment. Determining the ripeness of the capsule in the field can also be difficult and requires some experience with green capsule culture usually being undertaken weeks or a month or more before maturity. One way of judging capsule maturity is to squeeze the capsules and test the resistance and springiness—with green capsules being more resistant. However, knowledge of maturation duration and the time of year that maturation occurs are the most useful indicators of when a green capsule is at an advanced enough stage to be ready for sowing (see Case Study 4.4).

Procedure 4.6

Sowing seed from an immature capsule **(Figure 4.13)**

1. Cut the capsule from the orchid using a sterilized scalpel blade. For each capsule, use a new blade to prevent virus contamination. Capsules are best sown immediately, but may be stored for a couple of days before sowing, if necessary.
2. Trim the remaining dried flower parts.
3. Scrub the capsule carefully with a soft toothbrush and some detergent.
4. Soak the capsule in bleach (1% sodium hypochlorite) for 10 minutes.
5. In a laminar flow, remove the capsule from the bleach with sterile forceps and place in an empty sterile Petri dish to use as a cutting surface.

6. Holding the capsule with forceps, dip into 100% ethanol and pass the capsule through the flame of the Bunsen burner just enough to ignite the ethanol. Repeat twice, waiting for the flames to subside and the capsule and forceps to cool between flaming.
7. On a fresh Petri dish, use a scalpel to cut the capsule in half longitudinally.
8. Gently lift the capsule and hold upside down over a Petri plate containing germination media. Seed should fall from the capsule—give it a tap or scrape if necessary. Spread evenly and thinly over the plate or a number of plates. Repeat with the second half.
9. Generally, the most mature seed occurs in the center of the capsule so it can be useful to identify where in the capsule the seed was sourced. The capsule can be cut into multiple sections if that facilitates dispersal.

Figure 4.13 Sowing of an immature (green) seed from capsule

Case Study 4.4

Germination of threatened Asiatic Cypripedium *species (see also Case Study 6.2)*

Yung-I Lee
National Museum of Natural Science, Department of Biology, Taiwan

Cypripedium species are of high value in the ornamental horticultural trade with several species threatened because of habitat loss and over-collection. Thus, having a reliable protocol for producing seedlings is beneficial for conservation and commercial development. Unlike epiphytic orchids, seed germination of *Cypripedium* presents problems, as seed dormancy must be broken, and there are more stringent requirements for nutrient composition and cultural conditions (Rasmussen 1995). In comparison to the North American *Cypripedium* species (Light and MacConaill 1990; Steele 1996), less information on seed germination is available for the Asiatic *Cypripedium* species (Figure 4.14a). In this case study, we report scarification treatments of mature seeds and the optimum timing for seed collection to maximize germination

Figure 4.14 (a) *Cypripedium formosanum*; (b) *Cypripedium formosanum* protocorms; (c) *Cypripedium formosanum* germination demonstrated with a light micrograph showing an early globular embryo with a single-celled suspensor (S) of *C. formosanum* at 90 DAP. IT = inner testa; OT = outer testa. Scale bar = 20 μm. (d) Light micrograph showing a longitudinal section through a mature seed of *C. formosanum* at 210 DAP, enveloped by inner (arrowhead) and outer (arrow) testa. Scale bar = 40 μm. (e) SEM micrograph showing testa erosion (arrows) after treatment with hypochlorite solution. Scale bar = 50 μm (photos: Yung-I Lee).

of some Asiatic *Cypripedium* species. The procedures have been successful in producing thousands of seedlings.

Mature seeds of *Cypripedium* are more difficult to germinate than immature (green pod) seeds. As seeds mature, embryos are enclosed by an intact testa with strong hydrophobicity. Furthermore, the presence of inhibitory substances—for example, abscisic acid—have been detected in mature seeds (Lee et al. 2015). To break dormancy and improve germination, mature seeds were treated with chemical or mechanical scarification. Different hypochlorite formulations—that is, calcium or sodium hypochlorite—provide surface sterilization, as well as increasing the permeability of the testa and to leach/oxidize dormancy-inducing compounds from seeds.

Treatment with hypochlorite solutions (1% available chlorine) for 30–60 minutes improves germination of *C. lentiginosum*, *C. forrestii*, and *C. macranthos* (Miyoshi and Mii 1998). For seeds with a hardened testa—for example, *C. formosanum*—treatment with a 1% sulfuric acid solution for 10–30 minutes, with seeds then transferred to a 1% hypochlorite solution for 20 minutes, enhances germination (Figure 4.14b). *Cypripedium plectrochilum* is one of the most difficult species to germinate from mature seed because of its extremely thick inner testa. Even after treatment with 1% sulfuric acid solution for 60 minutes, germination is just 0.1%. In contrast, *C. debile* is easy to germinate in excess of 70% from mature seeds without any treatments (Hsu and Lee 2012). Unlike many *Cypripedium* species, *C. debile* does not possess a substantial testa and less hydrophobicity is exhibited.

The culture of immature seeds (green pod culture) is an alternative way to optimize germination without any pretreatment procedures. Since metabolic/mitotic activity is not arrested, the embryo continues to develop on media before induction of dormancy. The drawbacks of using immature seeds are that seeds have to be sown quickly after collection and long-term storage of immature seeds is not feasible, though cryogenic approaches could be possible. In addition, the optimum time for seed harvest may vary from one species to another, and indeed the time frame may change according to climate, seasonal conditions, and maternal influences. Usually, good germination can be obtained when seeds are harvested at the early globular stage (Figure 4.14c, d, e). Therefore, in using immature seed, a precise time frame in seed development should be investigated for each species. In *C. formosanum*, good germination is obtained with seeds collected at 90–105 days after pollination (DAP) (Lee et al. 2005); in *C. lentiginosum*, seeds collected at 90–120 DAP germinate well; in *C. macranthos*, seeds collected at 42 DAP germinate well (Zhang et al. 2013); and in *C. debile*, optimum germination is found as seeds approached maturity (120–150 DAP).

5

ASSESSING MYCORRHIZA IN SOIL

5.1 BACKGROUND

The relationship of orchids with their fungal partners is one of the more intriguing and complex associations involving plant mutualism with soil organisms—ranging from facultative to obligate. The study of orchid seed and germination *in situ* is significantly impeded by their dust-like seed and the difficulty of locating seed and seedlings in the wild. For terrestrial taxa, germination stages remain underground for several weeks, months, or even years before emerging above ground. Most of our knowledge of orchid fungi is based on *in vitro* germination studies using fungal cultures isolated from mature plants (Rasmussen and Whigham 1993). However, *in vitro* syntheses may not reflect the ecological and biological processes involved in the establishment process in nature.

Techniques to assess the mycorrhizal presence in wild sites provide useful guidance as to the role of mycorrhiza in orchid seed germination under natural conditions—how it may influence orchid distribution and for identifying suitable sites for orchid translocation. The approaches used involve direct assay of soil or sieved organic matter, or *in situ* baiting of habitat soils using retrievable packets of orchid seed (Brundrett et al. 2003; Masuhara and Katsuya 1994; Rasmussen and Whigham 1993). The difficulties associated with these methods include the relocation and retrieval of seeds following sowing, suitably positioning the seed to ensure close contact with soil and mycorrhizal fungus if present, and seasonal variation in environmental conditions such as moisture and temperature that can dry or overheat seed.

The techniques outlined in this chapter provide essential and ecologically relevant information for conservation practitioners and an experimental approach to determine ecological specialization of the mycorrhizal associations, mycorrhizal distribution in substrates, recruitment limitations, and niche exploration and competition. This chapter has two major components: first, an explanation of how these methods can be used for understanding ecological specialization in the orchid mycorrhizal interaction, including recruitment limitations at differing ecological scales; and second, identifying niche capacity for plant establishment and mycorrhizal *hotspots* at which to focus conservation translocations.

5.2 MYCORRHIZAL SPECIFICITY IN SOIL

Experimental approaches for assessing specificity and compatibility of mycorrhizal fungi are usually based on *in vitro* seed germination and show varying levels of germination specificity of the mycorrhiza (Hadley 1982; Masuhara and Katsuya 1994; Warcup 1971; Warcup 1981; Warcup and Talbot 1967; Zettler and Hofer 1998). The term *physiological specificity* or *potential specificity* applies to the associations observed under these *in vitro* conditions and should not be assumed to reflect levels of ecological efficacy under field situations (Batty et al. 2002; Rasmussen 2002). Orchid-mycorrhizal associations occurring in defined orchid habitats under field conditions are termed *ecological specificity* and may more accurately reflect the symbiotic association between plant and fungus (Harley and Smith 1983; Masuhara and Katsuya 1994).

Under natural field conditions, some orchids (Perkins and McGee 1995) may show greater specificity for fungal partners than under laboratory conditions. Batty et al. (2001) investigated the concept of ecological specificity by *baiting* for orchid-mycorrhizal fungi in natural populations of the terrestrial orchid *Caladenia arenicola*, to determine patterns of mycorrhizal distribution in the wild. Retrieved seed baits showed that germination increased in the vicinity of mature *C. arenicola* plants and there existed additional capacity for further orchid recruitment (Batty et al. 2001). A significant modification to this approach was the development of *in situ* and *ex situ* seed germination methods to simultaneously investigate the germination of different terrestrial orchids in the same soil sample or micro-niche (Brundrett et al. 2003). This study found that mycorrhizal fungi compatible with terrestrial orchids occurred most frequently in the organic matter and leaf litter in the habitat soil.

5.3 RECRUITMENT LIMITATION

Recruitment limitation in plants at the local (e.g., a single occupied habitat) and regional (the extent of a species habitat range) scale is reflected by the population size at a given time and the proportion of occupied sites across the landscape (Münzbergová and Herben 2005). Seed arriving at a site represents the first major limit on orchid distribution followed by germination and seedling establishment (which depends upon suitable environmental and mycorrhizal conditions). At the regional scale, if the limitation is due to carrying capacity where all suitable habitats in the landscape are occupied, the only possible way of increasing the number (or habitat range) of populations is by creating new habitats. However, restoring habitats to enable orchid establishment is an area of restoration ecology that requires further research, particularly since the specific habitat requirements of orchid mycorrhiza are often poorly understood (see Chapter 7). Altered ecological conditions such as changed fire regimes, weeds, soil disturbance, and altered vegetation structure, may mitigate against mycorrhizal competency that will ultimately limit orchid recruitment opportunity (Swarts et al. 2010).

Direct use of seed in orchid establishment and translocation uses two approaches: seed augmentation and seed introduction (Turnbull et al. 2000). In seed augmentation, seeds are sown to a site containing the parent orchid species where the abiotic conditions of substrate, microclimate, etc., are conducive to recruitment. Seed introduction involves sowing seed to sites where their absence may reflect unsuitable regeneration conditions, so that no amount of seed addition will lead to plant establishment. Recruitment following introduction experiments indicates that niches suitable for orchid recruitment exist, but that plant absence may be more of a case of a seed dispersal limitation (Turnbull et al. 2000).

5.4 DISTRIBUTION OF MYCORRHIZA IN SOILS

The spatial distribution of orchid mycorrhiza in soils is largely unknown. Organic matter, which is a primary source for mycotrophic nutrition, can be patchily distributed in soil (Brundrett and Abbott 1995) and may therefore limit fungal distribution. Inoculum potential is the amount of fungal material present in sufficient quantity to sustain germination and seedling growth (Garrett 1956). Our knowledge of how the inoculum potential of the soil influences orchid distribution is increasing through methods such as quantitative polymerase chain reaction (PCR) (see Chapter 9) and seed baiting studies, described in the upcoming text. Competition for mycelial contact occurs in *in vitro* cultures between orchid siblings (Alexander and Hadley 1983; Rasmussen et al. 1989; Tsutsui and Tomita 1989). It is therefore feasible that orchid seedlings in nature may compete for access to available mycorrhiza, though the carrying capacity of orchid mycorrhiza in natural conditions is poorly understood.

Clearly, for orchids with a requirement for a specific fungal partner, the distribution of the fungus will directly influence the distribution of the orchid. A better understanding of how and where orchid mycorrhizal fungi are distributed in the soil is important for mapping where and when to translocate orchids. The seed burial technique originally devised by Rasmussen and Whigham (1993) allows the distribution of effective orchid mycorrhizal fungi to be assessed *in situ*.

Early work by Curtis (1939) and Harvais and Hadley (1967) suggests that fungi associated with orchids are likely to be more widespread in the landscape than their hosts. In more recent studies, detailed *in situ* baiting has shown that there are more orchid-supportive micro-sites than there are orchid plants, and therefore landscapes could potentially support a greater abundance of orchids (Bailarote et al. 2012; Jacquemyn et al. 2011; Phillips et al. 2011). Strains of fungi known to associate with orchids have been isolated from soil in areas where orchids are absent (Warcup and Talbot 1967). The mycorrhizal fungi *Ceratobasidium cornigerum* and *Tulasnella calospora* that regularly associate with a wide range of orchid species appear to have a worldwide distribution, indicating that orchid recruitment capability is limited by other abiotic factors (which, in turn, may limit the germination efficacy of the mycorrhiza). Davis et al. (2015) demonstrated continental scale distribution of a *Sebacina* species which associated with the widespread *Pheladenia deformis*. Perkins et al. (1995) observed that the roots of *Microtis parviflora* hosted a range of combinations of mycorrhizal fungi across three broadly distributed sites, indicating that other factors, as well as host specificity, are important in determining which fungi were mycorrhizal with *M. parviflora* at any particular site.

5.5 NICHE EXPLOITATION AND COMPETITION

Once an orchid is established at a site through the arrival of seed, followed by germination and sustained growth and development in the presence of a compatible and efficacious mycorrhiza, exploitation of that niche may occur (Dixon 1991). Orchids, through growth of roots and *farming* of mycorrhiza, can increase the level of local mycorrhizal inoculum in soil and, therefore, the ability of the local site to support germination and establishment of orchid seedlings. In this way, a local symbiont for one species may outcompete other soil fungi or compatible symbionts of another orchid species. Many terrestrial orchids are well-known for promoting seedling germination in close proximity to the parent plant, thus increasing an orchid's ability to thoroughly exploit the micro-site around the plant (Figure 5.1). Seed burial and retrieval methods in a number of orchids including the achlorophyllous orchid *Neottis nidus-avis,* showed that germination occurred most frequently in plots containing adults of *N. nidus-avis* (McKendrick et al. 2002). The lesson from these studies is that often the best place for an orchid to germinate is where the mother plant has established—though many studies (see Case Study 7.3) show time and again that many more orchid niches exist if environmental conditions are created to enhance seed arrival and promote germination and development.

5.6 *IN SITU* SEED BAITING METHODS

Seed germination under natural field conditions has been attempted with varying levels of success. This has generally involved methods that allow for the burial and retrieval of orchid seed placed in porous retrievable packets. These have included packets of cotton gauze mesh (Masuhara and Katsuya 1994), Millipore screen envelopes (Perkins and McGee 1995), various slide frame designs (Batty et al. 2001; Diez 2007; Rasmussen and Whigham 1993), and nylon mesh packets (Brundrett et al. 2003; Swarts 2007).

More recently *in situ germination* has become known as *seed baiting* or *fungal baiting* as the purpose is generally to detect or *bait* for a mycorrhizal presence in the soil. Packets are

Figure 5.1 *Drakaea* orchid plant and seedlings (circled)

placed within the surface soil, usually beneath leaf litter, with the assumption that this is where seeds are most likely to aggregate and therefore afford the greatest chance of a successful encounter between the seed and a compatible mycorrhizal fungus. The type and distribution of organic material may also play a key role, with the highest fungal activity detected in coarse organic matter occurring within the top five cm of soil.

5.6.1 The Slide Frame Baiting Method

This seed burial technique devised by Rasmussen and Whigham (1993) allows for a direct, habitat-based assessment of the distribution of efficacious orchid mycorrhizal fungi (see Case Study 5.1). The technique has been successfully applied in a number of environments from temperate to tropical, with seedlings developing to the point of producing tubers (Batty et al. 2001). Frames can be made in advance and stored (for a few weeks) at room temperature until required (Figure 5.2). This technique is limited as it can only be used for one species at a time.

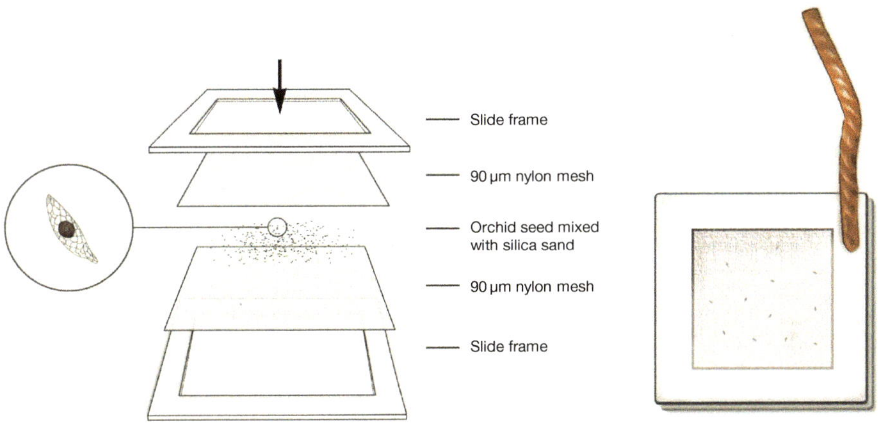

Figure 5.2 Slide frame method

Procedure 5.1

The slide frame method

1. Place the collected orchid seed (Chapter 4) with a small amount of sterile silica sand or a similar bulking agent to assist with evenly distributing the seed over a 40 mm × 40 mm piece of 90 µm nylon mesh. The diameter of the mesh is such that the holes are small enough to prevent the seed from falling through, while large enough to allow unimpeded movement of hyphae.

2. The quantity of seed placed in each slide frame will depend on the seed amount available, but approximately 100–300 seeds should be sufficient to ensure adequate soil screening for mycorrhiza.
3. Place a second piece of nylon mesh over the top and fix together a transparency slide frame on either side to lock the seed and mesh in place (or alternatively, heat seal margins if using plastic-based gauze).
4. To assist in locating the seed baits in the field, a ribbon or other locational material should be affixed to the slide frame.
5. Bury the slide frame in the top 50 mm soil organic-rich layer, leaving the locating ribbon clearly visible.
6. Supplementary watering with rain water may be required to maintain moist conditions for germination. The germination period depends upon the species; however, germinants remain visible for many months inside the packets—noting that seedlings cannot emerge from the enclosure.

Case Study 5.1

Using seed packets in orchid science and conservation

Dennis Whigham
Smithsonian Environmental Research Center, Edgewater, USA

Orchid seeds are tiny and working with them in field conditions was a problem until Hanne Rasmussen (Figure 5.3a) and Dennis Whigham published a simple method for enclosing the seeds in nylon plankton netting and placing the netting into an appropriate container before inserting them into field settings. The *seed packet* method is now widely used and the protocol has been adapted for use in terrestrial and now even epiphytic habitats. The technique serves many purposes. It provides a mechanism to determine how long seeds survive in nature, the timing of seed germination, and differences in germination in different habitats.

Following germination, protocorms (Figure 5.3b) are found in the seed packets; and they can be removed and used as a source of fungal material for isolation of the mycorrhizal fungus. The protocorm pelotons often can be cultured and used as a source of material for laboratory experiments, but they also can be identified using molecular techniques to determine if the fungi that are important for germination and protocorm growth are the same as the fungi that are isolated and identified from the roots of established plants. Even if the pelotons that are removed from protocorms in seed packets cannot be successfully cultured, DNA can be isolated and analyzed to determine the identity of the fungus. Primers, developed from fungi obtained from protocorms in seed packets or fungi isolated from the roots of established plants, can be used to determine if fungi are present in soils as well as quantifying the amount of fungus.

Here we give two examples of how the *seed packet* technique has been used. In one study seed packets were inserted into slide trays (Figure 5.3c), the spaces surrounding the seed packets were packed with different types and ages of ground wood (Figure 5.3d), and then the trays were placed into plots containing different types of substrates. The seed packets contained seeds of different species of orchids. This study demonstrated that substrate is important—especially

Figure 5.3 Slide frame method case study (a) Hanne Rasmussen placing a seed packet into soil; (b) Protocorms of the mycoheterotrophic *Corallorhiza odontorhiza* in a seed packet that had been in the field for one year; (c) site being prepared into which trays of seeds are being placed; (d) experimental array in the field. Trays with seed packets were buried in each small plot; the plots had ground wood of different tree species and the wood was chosen to represent different stages of decomposition. By using multiple trays in each plot, we were able to conduct the experiment for more than a year; (e) slide trays containing seed packets. The seed packets in each tray contain seeds of different species; (f) slide tray showing how ground and moistened wood was placed between the seed packets to assure substrate contact when the slide trays were placed into experimental plots in the forest (photos: Dennis Whigham).

the presence of soil organic matter. Seed packets have also been placed in spatial arrays to determine if germination rates are higher when seeds are placed near established plants (Figure 5.3e, f). By combining the placement of seed packets across the spatial arrays with soil sampling for the presence of fungi using primers, the study found that germination rates were higher closer to established plants and that the fungal patches were about 40–50 cm in size.

5.6.2 Multi-species *In Situ* Seed Baiting

This technique, which is based on the technique for a single orchid species as previously described, involves simultaneously baiting for a mycorrhizal presence with seeds of different orchid species.

Procedure 5.2

Multi-species in situ *seed baiting*

1. Develop a site sampling strategy that is statistically sound and addresses the key conservation questions being asked (e.g., what is the spatial distribution of efficacious mycorrhizal fungi?). This may involve setting up transects within sites to ensure analytical consistency.
2. Fold a strip of 90 µm nylon mesh in half and heat seal approximately 20 mm sections to make equal-sized compartments. The number of individual compartments will vary according to the experimental design, and the size of the compartments can also be customized (Figure 5.4a).

Figure 5.4 (a) Template of *in situ* seed baits showing nylon mesh heat sealed into compartments containing Millipore squares with orchid seed; (b) inserting seed baits into field sites at positions along a transect line.

3. The final compartment contains a locational tag (such as flagging tape) to assist in bait recovery.
4. First, sprinkle the seed over wet Millipore paper that has been cut into 10 mm × 10 mm squares to allow the seeds to imbibe moisture before the placement of baits *in situ*. Moistening allows accurate placement of the seed within each compartment and prevents seed lodging along the heat sealed seam.
5. Once the seed has been added, the top of each compartment is heat sealed to avoid loss. Keep moist (if sowing in the growing season) with a damp paper towel until placed in the soil.
6. Position the seed baits in the soil at the start of the growing season. The timing of this will be dependent on species and location.
7. At the site, make a 5 cm deep slit in the soil with a sharp spade or broad-bladed metal knife, being careful not to disturb the soil unnecessarily. Carefully embed the seed envelope vertically in the slit to just beneath the leaf litter layer (Figure 5.4b). Compress the soil to close over the envelope, leaving the locating tag above the leaf litter.
8. Supplementary watering may be required depending on seasonal conditions to maintain moist soil.
9. Seed baits are removed from the soil (depending on species) generally after three or four months for scoring of germination.

10. After retrieval, place the seed baits between layers of moist paper towels and place in a refrigerator to prevent the germinated protocorms from drying out (Figure 5.5). This preserves the size and shape of protocorms and allows pelotons to be extracted if required.
11. Score germination under a dissecting microscope according to the stages described in Chapter 4.

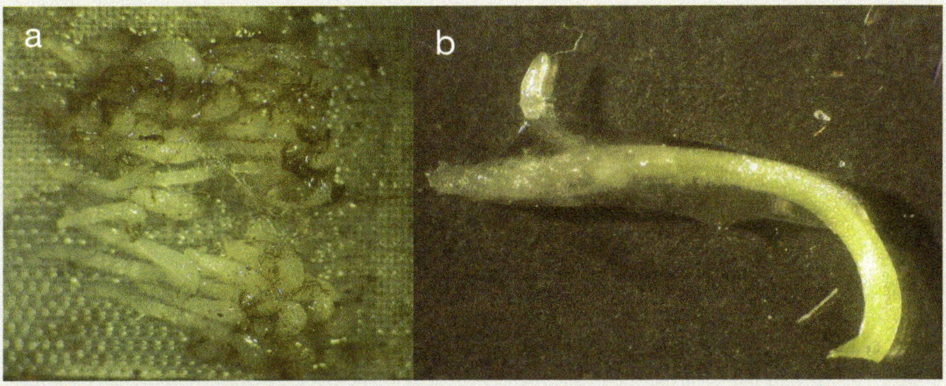

Figure 5.5 (a) Germinating protocorms in *in situ* baits; and (b) close up of germinated orchid seedling showing green leaf and root development even though the seedling was buried under leaf litter.

5.6.3 Seed Baiting for Epiphytic Orchids

The *slide* technique used in terrestrial baiting has been adapted for *in situ* baiting of tree branches and provides the same ability to assess if a tree surface can support orchid seed germination (Zi et al. 2014). The most effective technique involves using 2 × 2 cm to 5 × 5 cm squares of pliable plastic gauze, in which seed is incorporated with a small amount of finely chopped sphagnum moss; however, see Case Study 5.2 for an alternative method. The gauze square is stapled to the substrate with a staple in the middle as well, to ensure good contact with the bark substrate. Then moisten—noting that the gauze should be kept moist during the incubation period (4–8 weeks). It is best to install these gauze squares when the rainy season has just commenced, which is coincident with the time when epiphytic orchids germinate in the wild.

Case Study 5.2

Effectiveness of a new cultivation model in restoration of wild orchids used as traditional Chinese medicine: experiments from tropical Yunnan, China

Jiang-Yun Gao
State Key Laboratory for Conservation and Utilization of Bio-resources,
Lab of Ecology and Evolutionary Biology, Yunnan University, China

More than a third of Chinese native orchids are used as traditional Chinese medicine including many species of *Dendrobium*; and consequently, these orchids have been exploited to the level of local extirpation. Although commercial cultivation of medicinal *Dendrobium* has developed rapidly in recent years, this has not alleviated the pressure on wild harvesting. Our work shows a lack of engagement with local stakeholders—the farmers and town dwellers involved in wild collection are not involved in the commercial production of medicinal orchids. In addition, it is commonly believed that wild plants have superior medicinal qualities than cultivated products.

The restoration-friendly cultivation model we have developed attempts to address both of these issues. Planting targeted species in natural settings using local people is a new conservation approach for these epiphytic medicinal orchids, especially *Dendrobium* species. We set out to test this model in the medicinally valuable species, *D. aphyllum* and *D. devonianum*, using seed baiting techniques. We sowed seeds directly on tree trunks in natural forests and traditional tea gardens in 4 cm × 6 cm nylon packets with 45 μm diameter holes (Figure 5.6a).

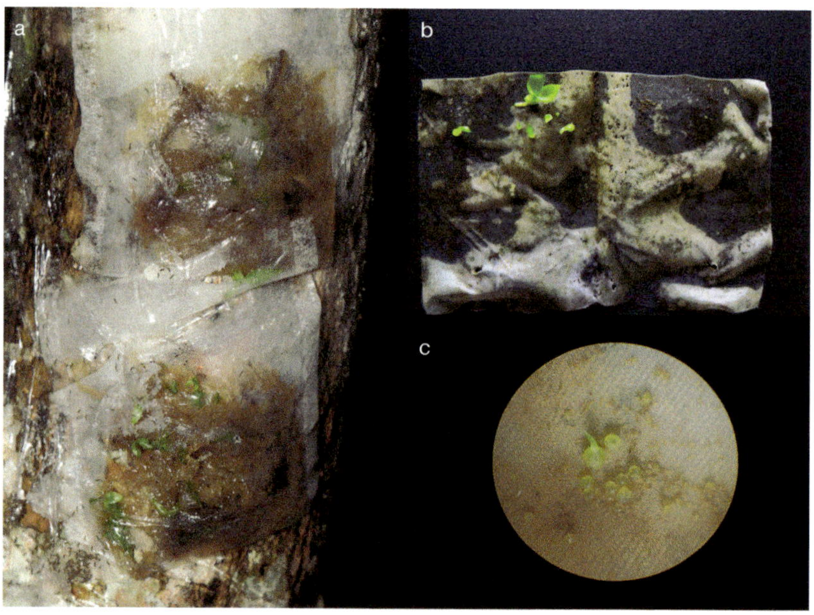

Figure 5.6 (a) *In situ* seed baits fixed to trees in the Xishuangbanna Tropical Botanical Garden, Yunnan, China; (b) *Dendrobium* protocorms in harvested *in situ* baits; and (c) *Dendrobium* protocorms from *in vitro* germination trials using mycorrhizal fungi isolated from *in situ* baits (photos: Jiang-Yun Gao).

The packets are porous to allow fungal hyphae to enter, but capable of retaining seeds inside the packet. Seed packets were secured onto the tree bark near the adult roots, and then covered with a thin layer of moss to prevent desiccation while allowing light to penetrate. Seeds started to germinate in six weeks, and seedlings developed successfully (Figure 5.6b). Symbiotic fungi were then isolated from protocorms of the germinating seeds. Using standard *in vitro* germination methods, the fungi were experimentally tested to be effective for seed germination and seedling development (Figure 5.6c).

Our experiments show that restoration-friendly cultivation could be an effective way to alleviate rural poverty and protect wild medicinal orchids (Liu et al. 2015; Zi et al. 2014). Equally significant, the symbiotic seed germination methods that we developed can be used for reintroduction of endangered orchids, including medicinally exploited taxa elsewhere.

5.6.4 Seed Germination in Retrieved Soil

This technique allows non-destructive observation of seed germination using field-collected soil samples or potting mix inoculated with the appropriate mycorrhizal fungus. Known as *ex situ* germination, this approach is a modification of a method devised to observe germination of fungal spores in contact with the soil solution (Brundrett and Juniper 1995; Brundrett et al. 2003). This is an effective method for observing germination out of season, as dry soil or frozen soils can be moistened or warmed to activate dormant fungi. The method also allows germination to be observed in the laboratory under conditions closer to natural situations than occur in sterile cultures. It does, however, rely on sieving coarse organic layer, which is considered to have the greatest potential for greatest mycorrhizal activity.

A disadvantage with this approach is that under the high humidity of the incubation vessel, soil organisms such as fungus gnats, mites, springtails, and nematodes, as well as other saprophytic fungi and slime molds, are active and may possibly compete with orchid fungi.

Germinating orchids via this method allows periodic, non-destructive observations and/or harvesting of developing protocorms. Protocorms can be used as an inoculum source, and the pelotons within can be isolated to identify the mycorrhizal fungus. Protocorm development with this method is similar morphologically to that of orchids germinated in sterile symbiotic culture.

Procedure 5.3

***Preparation of* ex situ *germination plates* (Figure 5.7)**

1. Develop a soil sampling strategy that is statistically sound and addresses the key conservation questions for the species or the sampling area.
2. Collect soil from field sites during the inactive fungal phase (summer or winter, if frozen). The soil is dried and sieved to remove the inorganic and organic components less likely to contain mycorrhizal fungi. The soil can be collected during the growing

season, however, sieving may not be possible due to being too wet, and disruption to mycorrhizal networks may lower inoculum potential.

3. To collect the soil, remove larger pieces of leaf litter, rocks, and organic debris. With a trowel, dig to a depth of approximately 5 cm from a 10 cm × 10 cm area.
4. In a clean, open workspace, break up the soil and pass through a 2 mm sieve, in stages, to retain the coarse organic matter.
5. Place a 5 mm deep layer of the coarse organic matter into 10 cm × 10 cm incubation dishes and moisten with water. The organic content may at first be hydrophobic. Allow to soak overnight and then remove excess water.
6. Cover the organic matter evenly with a thin layer of white, sterile silica sand, ensuring it is well moistened but not saturated.
7. Lay a 10 cm × 10 cm square, 90 mm nylon gauze over the top of the sand layer.

Figure 5.7 *Ex situ* baiting setup

8. Cut Millipore filter/membrane paper into 10 mm squares, lay side-by-side on a plate, and moisten with sterile water.
9. Sprinkle orchid seed over the Millipore squares and, with forceps, place in position over the nylon mesh. Here, multiple species can be used, or replicates of the same species, or a combination of both. The seeds do not need to be sterilized before use.
10. Cover the dishes, and incubate in the dark at a temperature matching the *in situ* conditions experienced by the species.
11. Plates must be checked regularly to ensure adequate water content and to remove weeds if observed.
12. At weekly intervals, inspect the dishes under a dissecting microscope to observe the number and stages of seed germination (Figure 5.8).

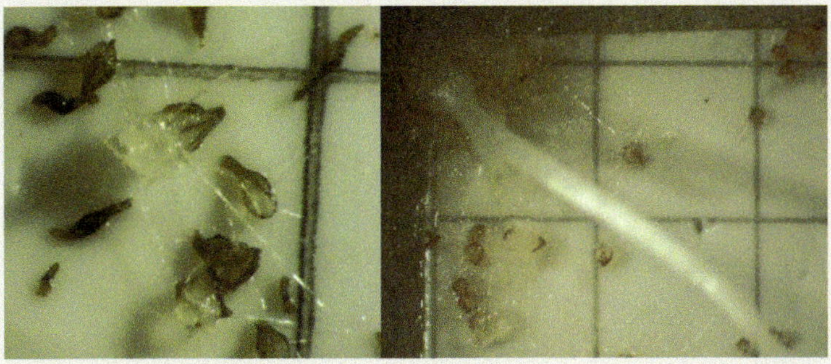

Figure 5.8 Germination of orchid seed in *ex situ* baits

6

ORCHID CULTURE FOR CONSERVATION

6.1 BACKGROUND

Orchid propagation for conservation is gaining increasing importance as an essential tool in developing effective orchid translocations—a fundamental and rewarding aspect of orchid conservation. Motivated by escalating threats and coupled with a decline in natural and suitable habitat, propagation science aims to successfully build or rebuild self-sustaining populations in safe sites linked to effective restoration approaches (see Chapter 7).

Orchid conservation science, as with many plant conservation approaches, is a continuum from theory to practice; obtaining critical knowledge from research for implementation by conservation managers/horticulturists in the field or glasshouse for the successful reintroduction of wild orchids. Scientists need effective and sound horticultural practices if they are to be able to translate their science into growing plants that survive and thrive under field conditions. Importantly, programs based on translocation often require long-term investment and strategic alliances with stakeholders (community, land managers, and researchers).

When choosing methods for orchid propagation, consideration must be given to the type, number, and availability of propagule sources. Vegetative propagation via division of plants, tubers, clonal spread, and *in vitro* cloning are effective methods for generating large numbers of plants—particularly for commercial purposes. However, for conservation or reintroduction programs, these methods may be limited in some instances due to lack of genetic diversity if clonal propagation is the only source of plant material and phytosanitary considerations such as accidental spreading of viruses. Propagation from seed provides a suitable alternative for orchid conservation programs, as this provides a ready means for ensuring genetic diversity (as long as seed is from outcrossed sources), in addition to the large number of seeds available to generate plants.

This chapter draws upon methods developed in Chapters 3 and 4 to outline requirements for effective orchid propagation for conservation. However, it is important to note that globally, enthusiasts, gardeners, nurseries, and community groups are often able to achieve outstanding propagation of species once thought intractable. To ensure you are up-to-date, join your local native orchid group to learn firsthand about techniques applicable to propagating your local orchids.

6.2 SYMBIOTIC ORCHID PROPAGATION

For most epiphytic orchids, propagation methods are well established—drawing upon the outstanding success of the commercial orchid horticultural industries. Terrestrial taxa have a higher requirement for associating with a mycorrhizal fungus for germination and establishment, which translates to more complexity when artificially propagated. For effective terrestrial orchid propagation, techniques have focused on conditions to improve the transfer of seedlings from an axenic culture to soil (Batty et al. 2006b) and optimization of tuberization and reemergence following dormancy (Batty et al. 2006a; Scade et al. 2006). There are surprisingly few cases where orchids, particularly terrestrial wild species, have been successfully translocated into the wild or reconstructed habitats with or without attendant mycorrhizal fungi (Ramsay and Dixon 2003; Yam et al. 2002), though techniques and approaches are being progressively developed (e.g., the genus *Cypripedium;* see Figure 6.1).

Currently, research on the establishment of terrestrial orchid seedlings in soil is limited, particularly for *in situ* reintroduction programs. On the other hand, it is well-known that most epiphytic orchids are easily propagated with high survival rates (commercial production techniques are well developed for many genera such as *Phalaenopsis, Dendrobium, Vanda, Cymbidium* and *Cattleya*) (see Figure 6.2). Germination for epiphytes is achieved on a sterile medium containing a carbon source and salts, and a mycorrhizal association is generally not required for continued growth.

Figure 6.1 *Cypripedium calceolus*

Figure 6.2 Commercial production of epiphytic orchids—seen here in Singapore where there are a wide variety of taxa, hybrids, and seedling flasks.

As outlined in the previous chapter, for symbiotic propagation of terrestrial orchids, some taxa are generalists, compatible with a range of mycorrhizal fungi, while others may depend upon just one or two fungal species (Ramsay et al. 1986; Warcup 1981). Orchid-mycorrhizal fungi can survive in the absence of the orchid, however, survival of terrestrial orchids is unlikely without a mycorrhizal association (Rasmussen 1995). For these reasons, terrestrial orchid propagation for conservation must consider the growth and habitat requirements of both the orchid and its mycorrhizal associate.

6.3 SEEDLING DEVELOPMENT

The germination continuum from seed to seedling: from seed to protocorm, emergence of a leafy shoot, root development, and finally full plant establishment (for geophytes these include production of tubers, rhizomes, and root clusters) are largely similar for terrestrial orchids. The germination of orchid seed can take from a week to more than one year depending on the species, and often this can be sporadic within a species or even a seed batch. The growth phases of terrestrial orchids are closely linked to the seasonal conditions in which they naturally grow (Figure 6.3). Therefore, the progression of these growth phases for propagation and the changes in mycorrhizal colonization and organ development must be matched to natural environmental stimuli.

The sequence of the development of seedlings beyond Stage 4 (Chapter 4), is largely dependent on whether the seedling is grown with or without mycorrhizal fungi and also on the life history of the orchid; for example, rhizomatous or tuberous; winter or summer active.

In temperate terrestrials, leaf emergence from dormant underground tubers occurs during the cooler winter months with the onset of winter rainfall (Figure 6.3). Flowering occurs during spring and early summer, either just before or immediately after the replacement of

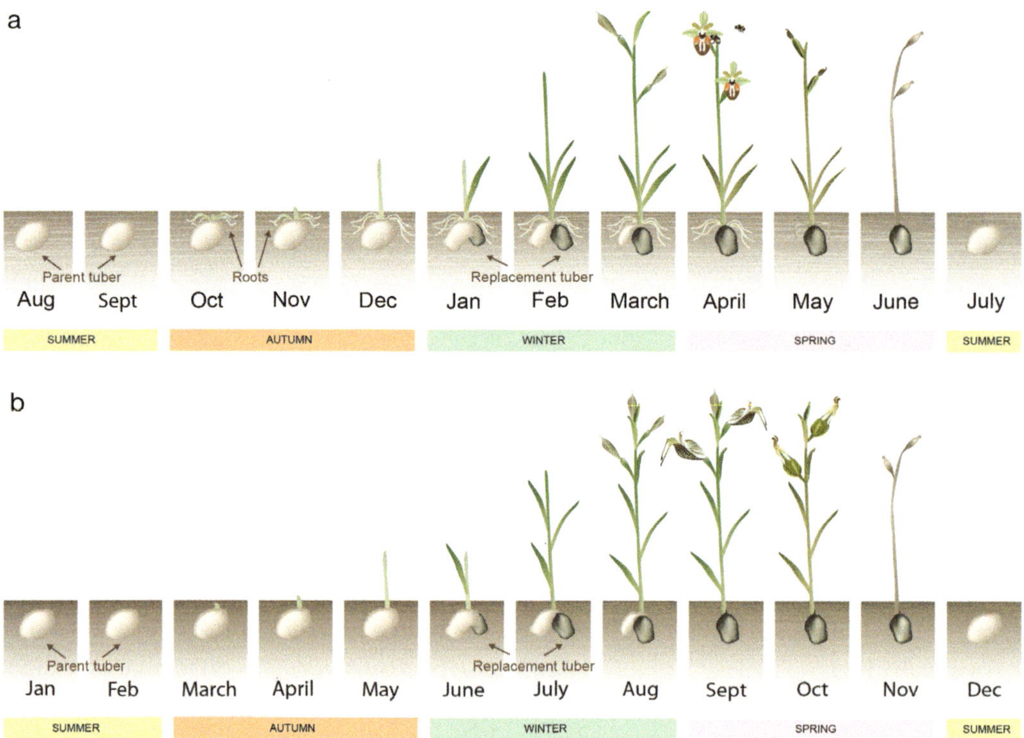

Figures 6.3 Phenology of growth and development in an herbaceous perennial orchid—(a) *Ophrys sphegodes* (northern hemisphere) and (b) *Pterostylis recurva* (southern hemisphere).

tubers. Plants enter their dormant phase as a quiescent tuber after their leaves and flowers dehisce generally with the onset of warm, dry summer conditions. For most of these taxa, sprouting occurs with the onset of moist cool conditions with a quick transition from the protocorm stage to leafy seedling. If the seedling survives the growing season—either in the glasshouse or the field—then a perennating structure (e.g., tuber, rhizome, or root cluster) with sufficient storage reserves has to be produced to survive the dormant period.

For cool climate species, such as many northern hemisphere orchids, the growing season is confined to a short period during spring with flowering in summer (some taxa can have a winter green period, though leaf growth is greatest in spring). To initiate the next stages of growth post germination (tuber development, breaking bud dormancy, etc.), protocorms from these taxa require cooling/vernalization of 4–5°C for up to three months. This also allows for the synchronization of natural leaf emergence and flower initiation for these orchids.

Tropical terrestrial taxa usually respond to the onset of the rainy season to break dormancy and therefore germination, unlike temperate taxa, usually requires warmer incubation temperatures (20–25°C). These species grow in warm, moist conditions with dormancy occurring in concert with a drop in temperature and drying of soils. Most tropical terrestrial species are fast-growing with leaf emergence and flowering occurring in some taxa within a few months of rainfall.

For the purposes of orchid conservation, the end-point to orchid propagation is the establishment of self-sustaining populations under natural conditions in the wild or restored habitats (Figure 6.4). An *ex situ* collection of orchids can also be used for seed farming (provided outbred seed is generated) to enable the building of seed banks for long-term protection of species.

Watchpoint: The genetic consequences of artificial propagation of orchids needs to be researched to ensure translocated plants are representative of the genetic diversity found in wild populations.

Figure 6.4 Terrestrial orchids in their natural habitat—(a) *Caladenia longicauda* (photo: Eric Swarts); (b) *Dactylorhiza praetermissa* (photo: Eric Swarts); and (c) *Dactylorhiza sambucina* (photo: Julienne Schiebold).

Case Study 6.1

An integrated approach to long-term orchid conservation in Africa

by Vincent Droissart[1] and Tariq Stévart[2]

In Africa, many orchids encountered in the field lack flowers and are therefore difficult to identify. To overcome that difficulty, a shadehouse cultivation system has been developed in West and Central Africa and in Madagascar to obtain fertile, identifiable material of orchid taxa (Figure 6.5a, b). These living collections are primarily used to conduct taxonomic studies. Since 1997, they have enabled the discovery of a new genus and more than 50 new species; and have also provided critical information for biogeographic and ecological studies, while concurrently assisting with the training of three Ph.D. and several Masters students.

The largest shadehouse of this network was established in Cameroon in 2004. Over the last 12 years, more than 6,500 living specimens representing around 250 species have been cultivated in Yaoundé, and have produced about 7,000 fluid-preserved specimens for identification. In

Figure 6.5 (a) *Cribbia pendula*; (b) *Ossiculum aurantiacum*; (c) comparative seed biology is undertaken for long-term storage; (d) controlled pollination; and (e) microscopic investigation of mature orchid fruits to determine seed quality (photos: Vincent Droissart).

[1]Institut de Recherche pour le Dévelopement (IRD), France

[2]Missouri Botanical Garden, St. Louis, MO, USA

November 2015, the National Geographic Society and the American Orchid Society sponsored a shadehouse that served for the establishment of the first orchid seed bank of tropical Africa. This undertaking aims to document the morphology and the viability, and to bank viable seeds, of 25% of the orchid diversity of Central Africa (about 150 species)—focusing as much as possible on the most threatened taxa.

A local technician was trained to (1) make controlled pollinations, (2) harvest mature fruits, (3) document morphology and seed viability (using the tetrazolium test), and (4) condition seed for long-term conservation in a freezer (−20°C) (Figure 6.5c).

A total of 397 manual pollination events (Figure 6.5d) have yielded 320 mature fruits and seeds from 122 epiphytic and 3 terrestrial orchid species, representing 31 genera in six tribes of Orchidaceae (Figure 6.5e). Fruit maturation time ranged from 45 days (*Tridactyle tridactylites* (Rolfe) Schltr.) to 452 days (*Vanilla africana* Lindl.). The smallest mature fruit (*Bolusiella zenkeri* (Kraenzl.) Schltr.) measured just 3.3 mm in length, whereas the largest (*Vanilla ochyrae* Szlach. & Olszewski) was 195.7 mm. Preliminary results showed that only 50% of the fruit harvested had viable seeds.

In parallel with this research on orchid seed conservation, we are currently assessing the International Union for Conservation of Nature (IUCN) conservation status of orchid species endemic to Atlantic Central Africa (ACA), which will help focus our conservation efforts on the most endangered African orchids. Preliminary conservation assessments of 175 taxa endemic to ACA reveal that most species are threatened: one might be extinct in the wild (*Angraecopsis dolabriformis* (Rolfe) Schltr.), 23 species are *Critically Endangered* (CR), 78 *Endangered* (EN), 34 *Vulnerable* (VU), 9 *Near Threatened* (NT), and only 30 were assessed as of *Least Concern* (LC). We have produced seeds for 10.3% of the species assessed as threatened (CR, EN, VU) based on application of the IUCN Red List categories and criteria.

While the current project focuses on Cameroon, it has significant potential over the coming years to link with and support other living collections in our tropical African shadehouse network. A duplicate collection of all seeds produced will be conserved by two key partners, the Missouri Botanical Garden and Kew's Millennium Seed Bank, which holds the world's largest *ex situ* seed collection of wild species for plant conservation.

6.4 TRANSFER TO SOIL MIX

Practical cultivation of terrestrial orchids is considerably less developed than for epiphytic species as most terrestrial species are not in commercial large-scale production. This is even less so for orchid reintroduction programs, although this is slowly being reversed with the advent of major species reintroduction programs (see references in Reiter et al. 2016).

A number of studies have investigated factors limiting propagation of herbaceous terrestrial orchids and subsequent survival and growth of seedlings transferred to field sites (Batty et al. 2006b; McKendrick 1995; Scade et al. 2006). After seed germination, problems associated with orchid transfer to a potting mix under greenhouse conditions or to field sites include high levels of initial mortality after the removal from sterile culture along with the inability to develop perennating organs (tubers) capable of surviving the dormancy phase. The following methods provide approaches for enhancing tuberization success in terrestrial orchids.

6.4.1 Sand/Vermiculite over Agar Growth Containers

Seedlings reaching the green leaf stage in the germination plate (at least 10 mm in height) should be transferred into disposable, polycarbonate containers as an intermediate step to facilitate the transfer of seedlings to glasshouse conditions (Figure 6.6). This method provides an intermediate stage between the sterile environment and soil, where seedlings are grown in an axenic environment with controlled humidity (Batty et al. 2006b). Although labor and time intensive, this method is particularly suited to rare species or where seed numbers are limited, while providing an effective means for optimizing survival after the transfer to soil, but also maximizing seedling size to facilitate tuber development. Batty et al. (2006b) found that seedlings from a range of genera produce larger tubers using this approach that are more likely to survive the dormancy phase.

Protocols for symbiotic germination of orchid seed using fungi isolated from host taxa have been shown to be effective in the propagation of seedlings that are most likely to survive the transfer to soil (see previous chapter). For most species, seedlings should be propagated at least three months prior to their typical field growing season. This facilitates the transfer of seedlings to the glasshouse during the cooler months of the year (for temperate orchids) and aligns plant growth as closely as possible to natural growth phases. The germinated seedling carries with it the mycorrhizal inoculum, usually in a sufficient amount to ensure successful outgrowth of the mycorrhiza.

Figure 6.6 Vermiculite agar with orchid seedlings

Procedure 6.1

Sand/Vermiculite over agar growth containers (Figure 6.7)

- Prepare heat-proof containers with a 7–10 mm layer of 0.25% oatmeal agar (OMA) (Chapter 3) autoclaved for 20 mins at 121°C.
- Once the OMA has solidified, a 5–10 mm layer of double sterilized (121°C for 20 min; rest 24 hr; 121°C for 20 min) silica sand or vermiculite is placed over the agar under sterile conditions in a laminar flow. Moisture from the agar is allowed to permeate the sand or vermiculite prior to planting.

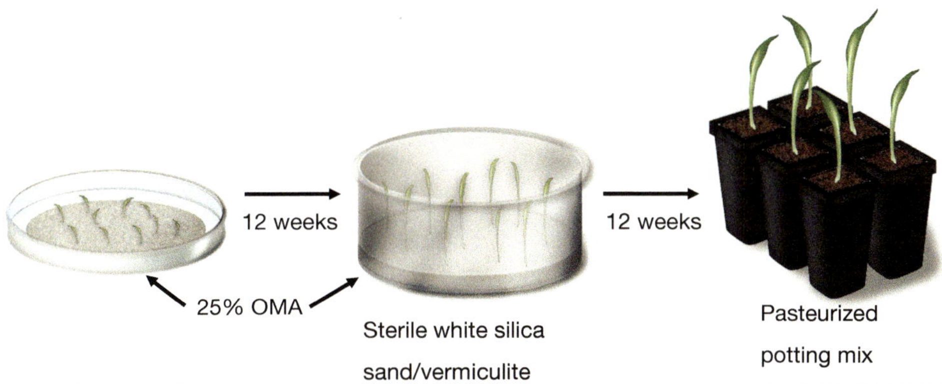

Figure 6.7 Seedling transfer from germination plate to growth containers

- Transfer symbiotically germinated seedlings into the sand/vermiculite using sterile forceps. It is important to transfer protocorms promptly after germination (when leaves are well developed; see Chapter 4), as seedling survival can be depressed if seedlings are oversized (transpirational loss or transplant trauma is exacerbated in large seedlings).
- Containers can be ventilated (if necessary) with Millipore discs placed over a hole in the lid.
- Containers with plantlets should be incubated at 18–22°C with a 16/8 hour light/dark regime, however, this may vary between species. Research on species from more extreme climatic regions is required to determine if different optimal temperatures or light regimes enhance the survival rate in the propagation process.

Watchpoint: Removal of advanced seedlings from culture plates, if done sufficiently early, can result in enhanced growth of remaining seedlings. Conversely, leaving precociously large seedlings on the original germination plate will often lead to suppression of growth in smaller seedlings.

6.4.2 Transfer to Potting Mix

Seedlings grown under *in vitro* conditions of high humidity and high availability of water have characteristically reduced cuticles and poorly functioning stomata. Therefore, the transition to conditions of lower humidity must involve gradual acclimatization. Following incubation for three months, growth containers can be placed in the glasshouse for one week before gradual removal of the lids. Seedlings are hardened-off for a further 5–10 days with the lids removed before the transfer to soil. At the soil transfer stage (Figure 6.8), seedlings should be washed thoroughly of adhering culture medium and then placed in a wetting agent (e.g., Envy) for ten minutes.

Washed seedlings are planted in potting mix in individual compartments (35 × 30 × 50 mm) of plastic seedling trays or forestry pots (Figure 6.9). Trays should then be covered with a clear plastic dome (humidity cover) with a variable controlled ventilation hole. Covers can be removed after 10 days, depending on how well the seedlings have adjusted to the conditions. Seedlings should be monitored for pests and fungal infection, particularly sciarid fly larvae that have a preference for protocorms of orchids.

If seedlings are grown via asymbiotic methods (using a nutrient rich media), it is at this stage where the seedling can either locate a mycorrhizal partner within the potting mix (or field site, see below), or the potting mix can be inoculated with the mycorrhizal fungus. A simple way to inoculate a soil mix is with mycorrhizal fungi grown in millet seed (see Section 6.6).

As soon as the seedlings have stabilized (leaf textures become more natural; leaves are self-supporting), they should be synchronized with the conditions experienced in the native habitat. This includes allowing seedlings to transition to a dormant state (usually by drying off the medium; raising or lowering temperatures). Tuber/rhizome development can be assessed once seedlings have entered dormancy (leaves have withered). Watering is recommenced when ambient temperatures are appropriate for the natural emergence of the species.

Figure 6.8 Orchid seedlings being removed from vermiculite agar containers ready for planting into potting mix

Figure 6.9 *Caladenia huegelii* seedlings transferred to a pasteurized potting mix in forestry pots under controlled glasshouse conditions

6.5 POT CULTIVATION

6.5.1 Requirements for Growing Terrestrial Orchids

Although growing structures for orchids can take a variety of sizes and forms, the basic requirements for orchids relate to providing adequate light, ventilation, and moisture. For many orchids, the controlled conditions of a glasshouse are unnecessary for optimal and successful growth, and a simple shade structure will suffice. Management of the growing area requires conditions that are hygienic, weed, pest, and disease free, and that potting mixes and fertilizer regimes are appropriate. Water quality is critical for many species particularly where plants are sensitive to pH and elevated levels of dissolved salts. Many of the conditions required (Table 6.1) to achieve optimum plant production can be manipulated according to the individual species needs. These conditions may include the composition; pH; moisture content, and temperature of the potting mix; fumigation and pasteurization practices; and the use of biocides and fungicides.

Table 6.1 Conditions and equipment for the propagation of terrestrial orchids for conservation

Cultivation	Requirements
Growing conditions	Light levels and ventilation must be adequate
	Cooling and/or heating may be required
	Raised benches are preferable to manage pest and disease issues
	Good drainage so mixes are open and not waterlogged
	A clean, weed-free environment is essential
Equipment	Clean, sterile pots
	Suitable potting mix
	System to pasteurize or fumigate mixes to remove harmful organisms
	Suitable fertilizers
	Water that is pure and uncontaminated
	Integrated pest management in place
	Controlled reticulation system (desirable)
	Capacity to control humidity

Case Study 6.2

Conservation through commercial production—a case of Chinese Cypripedium *species*

Holger Perner
Hengduan Mts. Biotechnology and Huanglong National Park, China

The genus *Cypripedium* is distributed over the Northern Hemisphere, except for Northern Africa. Populations, however, occur infrequently and can be rather rare, but within populations, the plants can be locally abundant. That makes them easy targets for commercial collectors who can destroy a population through digging within a day's work. *Cypripedium* species are in demand for horticulture since the nineteenth-century. Until the 1980s, the supply was exclusively provided by plants collected in the wild. But pioneers of artificial laboratory propagation from seed, like Carson Whitlow (USA), Werner Frosch (Germany), Svante Malmgren (Sweden), and William K. Steele (USA) have paved a road for commercial mass-propagation. Their work is based on scientific studies by Gaetan Harvais (Canada) and Johan Van Waes (Belgium).

A key aspect to propagating *Cypripedium* from seed is to overcome the physical and biochemical germination barrier of the inner seed coat, called carapace (derived from the inner integument) and the embryo cells at the suspensor end. The most effective way to achieve this is to sow immature (green pod) seeds (Figure 6.10a), which is for most species 8–9 weeks after pollination. At that stage, the carapace still consists of living cells that are permeable to water and nutrients, leading to germination (Figure 6.10b). When using ripe seeds, an aggressive treatment with hypochlorite (bleach) to chemically corrode the carapace enhances permeability in the same way. A medium low in mineral salts, providing nitrogen in organic

Figure 6.10 (a) Sowing unripe seeds from a surface sterilized green capsule; (b) germination with development of protocorms in constant darkness; (c) seedlings develop in constant darkness; (d) vernalization of deflasked seedlings at 2–4°C for 4–5 months; (e) planting out seedlings in a perlite mix in nursery beds at 2,960 m altitude; and (f) first flowers appear four years after planting (photos: Holger Perner).

form, is essential for seedling development (Figure 6.10c). With well-established growing techniques (Figure 6.10d, e), deflasked seedlings will reach flowering within 3–6 years and can be commercially produced in quantities up to hundreds of thousands (Figure 6.10f). Hengduan Mountains Biotechnology, Ltd. in Sichuan, China, has been undertaking the commercial production of *Cypripedium* using these approaches since 2003, and focuses on providing Chinese *Cypripedium* species for the international market, thus relieving the pressure of wild collection and also providing material for translocation programs.

6.5.2 Potting Mixes

Whereas epiphytic orchids can grow on a variety of substrates, mixes, and even in the air (many Vandaceous species are grown this way), growing terrestrial orchids relies on providing a suitable substrate for root growth and perennating organ development or replacement while preserving the mycorrhizal symbiont. Potting mixes for terrestrial orchids have in common a high proportion of organic content and an open, free-draining texture. Naturally this will vary between species, particularly as some occur in almost pure sand, while others grow in loams, organic mulch, or leaf litter. North American terrestrial orchids can be grown in mixes comprising sphagnum, perlite, vermiculite, and sand. To accurately match natural field conditions, it may be necessary to use soil from field sites or prepare a potting mix that mimics the texture, drainage, and water requirements of field sites—some experimentation may be required. Though not recommended, when using soil collected from field sites, be aware of phytohygiene issues and pasteurize if in doubt.

As there is variation in the requirements for the effective growth of orchid seedlings and mature plants, nurseries will generally create a potting mix comprising a range of soil mix components. These components include peat, sand or aggregate, vermiculite or perlite, composted tree bark, and other wood products that are combined in various proportions (Table 6.2) to make a *soil-less* potting mixture. Note: potting mixes must always be open, free draining, and contain organic matter for mycorrhizal fungi. The following is a brief description of the key components used in orchid potting mixes.

6.5.2.1 Peat

Peat is a major component of many potting mixtures providing an organic substrate for root growth. There are many variations in *peat* formulations used by nursery managers; and

Table 6.2 Composition of four different types of potting mixes

Mix	Components	Proportions	Orchid Species
Mix 1	Coarse sand	4.5 parts	General
	Loam	4.0 parts	
	Peat moss*	1.5 parts	
Mix 2	Coarse sand	2.0 parts	Temperate terrestrial taxa
	Rich loam	1.0 parts	
	Wood chips (well leached and composted)	1.0 parts	
	Composted leaf litter	1.0 parts	
Mix 3	Coarse sand	2.5 parts	General
	Coastal plain sand	5.0 parts	
	Loam	1.0 part	
	Peat moss	1.5 parts	
Mix 4	Loam	1 part	European species
	Gritty sand	2 parts	
	Fine pine bark	2 parts	
	Leaf mold (oak or beach)	1 part	

Note: the type of mix will depend upon the moisture and porosity requirements of an individual species. Consult with your local orchid group for advice on the appropriate mix combination for your particular orchid species.

*Ensure moss is sustainably sourced

these include sedge-based peats, shredded coconut, and more recently, coir peat derived from coir fiber bales. Peat derived from sphagnum moss has excellent physical and chemical balances, being low in available nutrients and pH. For container mixes, peat is often used as a 50% component with other ingredients such as vermiculite and perlite. Note: only peat from sustainable sources should be used.

6.5.2.2 *Sand or Aggregate*

These are added to assist drainage and add bulk density (weight) to the mix and reduce shrinkage in dry mixes during drying-out periods (dormancy).

6.5.2.3 *Vermiculite and Perlite*

These are relatively cheap and readily available components that bulk the potting mix, retain moisture, and ensure the mix is free-draining. Both materials are lightweight and prevent the potting mix from settling and compacting. Perlite lacks any buffering capacity and is nutritionally inert, while vermiculite has a higher buffering capacity and can provide Mg and K for seedling growth (Goh and Haynes 1977).

6.5.2.4 *Composted Tree Bark*

If using bark-based materials, the best is composted, finely mulched, hardwood bark. Hardwood barks can support mycorrhizal growth and suppress pathogens, however, ensure that the bark material has been adequately composted to remove excess tannins and other growth suppressing components.

6.5.2.5 *Composted Wood Products*

Composted sawdust and fine wood chips are used as natural substitutes for peat. Some types of wood can be phytotoxic, and preliminary testing of these products is essential prior to their use. Aged and composted wood products are essential. Softwood products can contain high levels of phytotoxic resins, so ensure that these materials are well composted before use, or resort to composted hardwood products.

6.5.2.6 *Composts*

Compost is manufactured by taking straw, wood products, and plant residues, and maintaining the mixture in a moist state to allow partial microbial decomposition (Hardy and Sivasithamparam 1989; Hoitink 1980). Often, nitrogen is added if wood materials are used to maintain a suitable nitrogen to phosphorus balance. Compost has a high water retention, is low weight, and mixes well with sand, soil, perlite, or vermiculite. Potting mixes based on composted hardwood bark fines have been found to suppress pathogens such as *Phytophthora* spp. (Hardy and Sivasithamparam 1991; Hoitink 1980). No animal wastes are used.

6.5.3 Pasteurizing Potting Mixes

Potting mixes are best pasteurized when they are warm and moist; however, peat can be steamed when it is almost dry. Pasteurization is generally carried out by soil steaming in open fields, greenhouses, or trailers. Steam treatment, when carried out at 70°C for 30 minutes, is essentially a partial disinfection, as it eliminates harmful microorganisms, allowing some beneficial microbes to remain (Baker and Roistacher 1957). It is important that much of the saprophytic microflora remains to revitalize the soil after cooling. An additional goal for nurseries that are cultivating mycorrhiza is to eradicate less efficient strains of mycorrhizal fungi, such as *Thelephora* spp., which can reduce the success of controlled inoculation programs.

Potting mixes with sphagnum peat, perlite, or vermiculite as major components often do not require pasteurization. Mixes containing sawdust (preferably composted), bark, raw humus, or soil should be pasteurized to prevent weed seed germination and reduce the potential threat from pathogens and other microorganisms. These could also act to limit or outcompete desired mycorrhizal fungi growth in the root.

A wide variety of steam pasteurization equipment is available, but may be expensive for small-scale production (Figure 6.11). Temperatures exceeding 70°C can be disadvantageous in potting mixes containing some soil types (particularly acidic soil) due to the release of toxic amounts of components including manganese (Sonneveld 1979).

Solar pasteurization is a cheaper though less reliable alternative to the use of steam, where clear polyethylene sheeting is used to cover or enclose moist potting mixes followed by exposure to direct sunlight for up to two weeks. Temperatures can rise to over 60°C under the polyethylene sheet and may be higher in the tropics. Solarization enables the continuous or repeated exposure of potting mix to a sublethal temperature which can either kill the majority of the pests, weed seeds, and pathogens, or weaken them so that they cannot survive (Porter and Merriman 1985; Stapleton et al. 1985).

Chemical sterilization with fumigants can be simple and effective, but are not desirable due to the elimination of beneficial microorganisms and safety concerns with handling often dangerous chemicals.

Figure 6.11 Steam pasteurizer trailer for use in cleansing pots and mixes

6.5.4 Growing Conditions

Note: your local orchid group (https://www.orchidmall.com/society.htm#sa) can provide locally accurate information on growing conditions—visit the following websites for further details:

- American Orchid Society: http://www.aos.org/orchids/orchid-care/orchids-101.aspx
- Hardy Orchid Society: http://www.hardyorchidsociety.org.uk/HOS%201012/Cultivation.html
- Australian Native Orchid Society: http://www.anosvic.org.au/ANOS_Vic_Merchandise.html
- South Africa Orchid Society: http://www.saoc.co.za/aerangis-exquisite-african-orchids.html
- The Orchid Society of South-East Asia: http://www.ossea.org.sg/index.php/publications/others
- Brazilian Orchids: http://www.delfinadearaujo.com/page2.htm

The conditions for growing many commercially propagated orchids are well-known and have been described elsewhere (Arditti 2009; Rasmussen 1995). While many of the conditions required for growing common species may be cross applied to threatened, rare, and unpopular species, important consideration must be given to soil moisture and watering, nutrient supply, and environmental conditions such as light and temperature.

- *Watering*: Watering regimes will vary between species so consideration must be given to natural growth cycles. As a rule of thumb, summer active orchids are maintained dry or just moist during winter, whereas winter/spring active orchids are maintained dry (with occasional wettings) during warm, summer conditions.
- *Nutrient supply*: The use of artificial fertilizers is generally not recommended for herbaceous terrestrial orchids with nutrients being provided via organic matter. Epiphytic taxa have a number of proprietary products available and regular application of diluted fertilizer is necessary to ensure good growth. Specific orchid commercial fertilizers are available, and if used, it is essential that fertilizers are accurately and uniformly applied to plants. Special attention to macronutrients such as manganese, boron, and zinc are known to facilitate plant growth. In one study on a common Australian terrestrial orchid, *Caladenia arenicola*, potassium was found to be linked to *in situ* plots where germination was highest (Batty et al. 2001).
- *Environmental conditions*: The greenhouse/growing environment should be monitored to ensure light and temperature conditions are favorable to plant growth. Consideration should be given to the environmental conditions that the orchids experience in their natural habitat. While for many species there is considerable tolerance in the range of conditions an orchid will thrive in culture, for others, more specialized conditions may be required. For example, the windswept helmet orchid *Nematocerus dienemum* which occurs on the sub-Antarctic Macquarie Island, is grown in a refrigerated shipping container to mimic conditions on the island (Figure 6.12).

Figure 6.12 (a) *Nematocerus dienemum* (photo: Peter Fehre); (b) *Nematocerus dienemum* growing in a refrigerated shipping container.

6.5.5 Disease and Pest Control

The propagation of orchids in nursery conditions creates opportunities for pests and diseases to flourish if not checked and maintained. Poor nursery hygiene and plant stress will exacerbate these problems. The management of pests and diseases is simplified if seedlings are grown in containers on benches or in raised beds (Figure 6.13). This reduces the direct exposure to soil pests and pathogens. Orchid growers need to consider appropriate strategies for controlling pests (Figure 6.14) and diseases (Figure 6.15) which include the interactions between the causal organism, the containerized mycorrhizal plant, and environmental factors. This may involve an overall integrated pest management strategy which can be adapted to specific growing regions or glasshouses. In some instances, chemicals may be

Figure 6.13 Growing orchids on benches under traditionally controlled greenhouse conditions

Figure 6.14 Yellow insect paper to ensnare pests

Figure 6.15 Diseased orchid leaves showing evidence of a striated virus. Plants with virus disease must be carefully removed, avoiding contact with healthy plants in the collection and destroyed by combustion.

required to control insect pests (insecticides) and nematodes (nematicides). Please note the material safety data sheet for each chemical and follow recommended safety procedures.

Viruses present a significant risk and challenge in orchid propagation facilities. Both epiphytic and terrestrial taxa carry and are impacted by viruses. If orchids are being grown under closed conditions such as in a greenhouse, then plants must be inspected for signs of streaking, blotching, or necrotic spots or lesions on the leaves. Malformed shoots and flowers are a sure sign of a virus, though many viruses are mildly symptomatic so it is easy to overlook an infection. Infected plants *must* be destroyed and monitoring of nearby plants undertaken to ensure cross-infection has not occurred.

6.5.6 Repotting of Terrestrial Orchids

Many terrestrial orchids increase in number each year through either cloning or self-seeding. Repotting herbaceous terrestrial (tuberous) orchids is most effectively done while the plants are dormant and requires preservation of some of the original potting mix as an inoculum source for the new mix. Potting media, if incorporating naturally sourced organic materials, often contain the required mycorrhizal fungus suitable for many root colonized orchids (for example, northern hemisphere taxa; southern hemisphere taxa such as *Thelymitra, Microtis,* and *Prasophyllum*). For rhizome and stem infected taxa, specific mycorrhizal inoculum is required or transfer of some of the original potting mix into the new potting mix.

Procedure 6.2

Repotting of orchid tubers **(Figure 6.16).**

1. Empty the pot and separate the tubers from the potting mix. A sieve may be required. For some species there may be a remnant sheath through which the new stem will grow. This should be preserved and replaced in the new pots.
2. Make up a potting mix (refer to Table 6.2) and fill half the pot (assuming a 15 cm pot).
3. Place the tubers with the eye or shoot upward in the mix. About 5 tubers are sufficient in a 15 cm pot, however, more can be added.

Figure 6.16 Orchid tubers that need replanting (photo: Richard Thompson)

4. Fill another quarter of the pot with the new mix.
5. For the remaining quarter, fill the pot with a 50:50 portion of new potting mix with the old mix.
6. Place a thin layer of organic mulch on top of the pot. Moisten lightly.
7. Keep the pot almost dry (occasional misting or gentle watering can assist in maintaining healthy tubers) until commencement of the growing season begins; then it should be watered and kept damp until the leaves emerge.
8. Water as required.

6.6 DIRECT SEEDING

Alternative methods to laboratory-based seed germination is to sow orchid seed directly to potting mix previously inoculated with a compatible mycorrhizal fungi or to pots containing mature plants. These methods are most suitable when large amounts of seed are available as adding seed to soil can be quite wasteful. Refer to Chapter 5 for *in situ* experimental germination methods.

6.6.1 Inoculating Potting Mix

Efficacious mycorrhizal fungi must first be isolated as described in Chapter 3 and cultured on millet seed or another inoculum carrier. This method still has an element of risk involved as the seeds must be compatible with the fungus inoculum, and the inoculum needs to be able to spread throughout the potting mix following incorporation of the inoculated millet seed.

Procedure 6.3

Preparing millet seed inoculum

1. Double sterilize 30 g of white millet seed (Figure 6.17a) in 30 ml of deionized water in glass jars in an autoclave (121°C for 20 minutes; rest 24 hr; 121°C for 20 minutes).
2. After each autoclave, shake the jars thoroughly to agitate the millet as sometimes the millet in the center of the jar remains non-sterile. In some cases, even a third autoclaving may be necessary to remove all the bacteria.
3. Allow the jars to cool.
4. In the laminar flow cabinet, cut the mycorrhizal culture (grown on potato dextrose agar (PDA)) in the petri dish into 0.5 cm blocks.

Figure 6.17 (a) Millet seed inoculum; (b) germination of orchid seed in potting mix containing millet seed inoculated with mycorrhizal fungi.

5. Add approximately one fourth of the petri dish into the glass jar with the millet.
6. Shake the jar to disperse the inoculum blocks through the millet seed.
7. Every few days, vigorously shake the jar to ensure the millet seed is free moving and to ensure the spread of the inoculum. Do not allow the millet seed to aggregate. If aggregation occurs, open the jar in a sterile environment and using a sterile knife agitate and break up clumps.
8. Discard the millet seed in the jar if it develops a wet or slimy (bacterial contaminants) or dirty appearance (fungal contaminants).
9. After 8 weeks, remove a random selection of individual millet seeds and plate onto Petri dishes containing PDA and outgrow to determine that the mycorrhiza is viable.

Once a pure (no fungal or bacterial contaminants) inoculum has been obtained and confirmed by replating onto PDA, the millet seed can be mixed in with the top 10 cm of the potting mix. Note: the potting mix should be a natural organic mix reminiscent of the conditions under which the seed would germinate naturally. For example, use leaf mulch taken from the forest floor for orchids that naturally grow in that area—ensuring that the mix is free-draining, disease free, and pasteurized before use.

Orchid seed can be added to the potting mix by combining the seed with sterile white sand (so you can see the brown seeds) and sprinkled over the soil surface using a salt shaker. Keeping the potting mix moist and placing pots under a humidity cover as previously described, or using a fogging apparatus will ensure good germination. Germination can take place within a matter of weeks, however, this depends on the species used. Protocorms from temperate terrestrials germinated using this method emerge within 8–12 weeks (Figure 6.17b). Seedlings can be moved when at a manageable stage, although it is preferable to leave seedlings till the onset of dormancy and transfer dormant tubers to individual pots. Ensure some original potting mix is transferred to guarantee adequate cross-inoculation of mycorrhizal fungi.

6.6.2 Mature Plants

A tried and proven method for propagating terrestrial taxa is to directly sow seed around an actively growing orchid plant of the same or closely related species (Figure 6.18). In general, if a mature plant is actively growing in a pot, it can be assumed there is a compatible mycorrhizal fungus available. Seed from that species can be sown around the emerging mature plant during the growing period using seed diluted in free flowing sand—sprinkle lightly, as seedling density may be high. Once seedlings have been established in the pots, they can be removed and transplanted into new pots—or preferably, moved as dormant tubers.

For most terrestrial orchids, the fungus associating with the parent plant is sufficient to support germination and seedling growth. However, in some cases, the fungus associated with the parent plant may be different to that required for germination—as found in the commercial production of *Gastrodia* in China.

Figure 6.18 Adult plants in pots with a range of juvenile seedling stages arising from direct sowing of seed

7

HERBACEOUS TERRESTRIAL ORCHID TRANSLOCATION

7.1 BACKGROUND

As global biodiversity is declining rapidly, action is urgently required to reverse the decline of endangered species and prevent extinction. Translocation, in its variety of forms, provides the opportunity to create self-sustaining populations while mitigating threatening processes facing orchid populations. However, for translocation to be successful, an in-depth understanding of existing threats, species ecological requirements, and habitat restoration is required (Reiter et al. 2016).

The main goal of plant translocation programs is to establish resilient, self-sustaining populations that have sufficient genetic resources to undergo adaptive evolutionary change (Guerrant Jr. 1996), thus enhancing prospects for species survival in the wild. For plants, the term *translocation* refers to any movement of living organisms from one place to another (International Union for Conservation of Nature (IUCN) 1987). Within the spectrum of conservation, there are a range of translocation types (Seddon 2010). Here we describe three that include both juvenile and mature plants (Reiter et al. 2016).

- *Introduction*: movement of a plant outside its historic known native range (also known as assisted migration, assisted colonization, or managed relocation).
- *Reintroduction*: intentional movement of a plant into a part of its native range from which it has become extirpated (also called reestablishment or out-planting). This may be part of the process of restoration or reconstruction of habitat. This can be further described as the movement of seed or plants propagated *ex situ* to the wild, either within the species' natural range or within its home range.
- *Transplanting*: the deliberate movement of wild individuals or populations from one part of their range to another (also referred to as salvage, restocking, or enhancement). This is usually completed as an emergency response to a development or land clearing, where the orchids would otherwise be destroyed.

Successful translocation remains a significant test for plant conservationists with increasing evidence that material translocated from *ex situ* collections is important for *conservation insurance* in the case of the extinction of wild populations. Note: the technological capacity to successfully propagate or translocate an orchid species should not be used to predicate the destruction of orchid habitat or loss of the species.

Herbaceous terrestrial orchids occur from tropical to cool temperate regions and represent some of the world's most endangered orchids. This chapter discusses the translocation of orchids with a seasonal dormancy, when the plant senesces to a dormant rhizome, tuber, or root tuber cluster. Approaches and techniques that can be applied to conservation translocations involving herbaceous perennial orchids are also presented.

7.2 REQUIREMENTS FOR SUCCESSFUL ORCHID TRANSLOCATION

Translocation programs need to consider the requirements for plant populations to persist in the short and long term. These include:

- Successful establishment of the translocated individuals
- Management and control of threats
- Sufficient numbers to avoid both demographic and environmental stochasticity
- Reproduction and natural recruitment
- Sufficient genetic diversity to retain evolutionary potential to adapt to long-term change

Translocation may be the only method for ensuring the survival of some threatened plant species that are represented by only a few populations. Despite the advantages of successful translocations, there are possible risks associated that must first be considered before the process is commenced. For example, in rare instances, mixing individuals from different populations may create outbreeding depression and a reduction in offspring fitness to adapt to changed environmental conditions. In the absence of strict phytohygiene standards, translocations may result in the accidental introduction of diseases or pathogens. Also, displacement or hybridization with other local species may occur if plants were introduced into a habitat where the species has not occurred previously. In addition, activities associated with translocation, such as disturbance to the soil and trampling, may have detrimental impacts on local vegetation.

Reiter et al. (2016) undertook a critical analysis of 74 published translocation studies involving 66 species of orchids. Each of the three translocation types was represented, with the majority being *introductions*, where orchids were planted outside their known home range. The analysis found that the time of post translocation monitoring ranged from 1–11 years, and the average survival rate was 66%, but ranged from 0–100%. A significant finding was that plants might survive until flowering, but then fail to set seed because of a lack of pollinators. Many of the studies reviewed did not relate the performance of the translocated orchids to wild populations, nor did they consider the critical ecological requirements for orchid survival, such as the presence of compatible mycorrhizal fungi or suitable pollinators. Reiter et al. (2016) concluded that although orchid translocations around the world are achieving rates of survival ahead of plant translocations in general, very few have resulted in self-sustaining populations. Considerable work still needs to be done to improve the survival rates of orchid translocations in the future.

7.3 PROPAGATION FOR TRANSLOCATION

Translocation programs where artificially propagated orchid seedlings are moved into new sites or used to supplement depleted populations, or the emergency transplanting of mature plants to new sites, are being used as tools in orchid conservation (Zettler and Mcinnis 1992). The programs rely on established principles relating to the importance of a *safe site* for reintroduction of orchids, where the presence of a compatible mycorrhizal fungus has been confirmed or is expected, and the specific plant pollinator is present (Batty et al. 2002).

Given the paucity of knowledge on the ecological requirements of both mycorrhizal fungi and pollinators, there is limited capacity to effectively restore a plant community that sustains and enhances these important agents in orchid development (Dixon and Hopper 1996). Ensuring the site selected for reintroduction matches the existing habitat in its vegetation characteristics, soil composition, underlying geology and topography, and hydrology is of critical importance for ensuring the long-term survival of reintroduced orchids. Mycorrhizal baiting using *in situ* or *ex situ* methods, as described in Chapter 6, are effective ways for determining mycorrhizal presence at recipient sites, though often these techniques underestimate the capacity for an orchid to be successfully translocated. In addition, baiting techniques allow for testing the persistence of resident or introduced mycorrhizal fungi (from millet seed inoculum—see Chapter 6) in successive years after the translocation. Ongoing monitoring of plant growth and development, including effective control of pests and diseases, is also critical for the success of recovery programs (Ramsay and Dixon 2003).

The combination of mycorrhizal, pollinator, and follow up monitoring of plant performance, provides ecological and logistical parameters that are likely to be important for influencing the long-term success of a recovery program of translocated orchids. Failure to consider these factors in the planning for orchid recovery means that such programs can represent little more than gardening (Hobbs and Cramer 2008).

Reintroduction programs should not be seen as a substitute for primary habitat conservation or community involvement in *in situ* management. Rather, artificial manipulation of plants in wild sites should only be undertaken in the context of enhancing the conservation status of protected habitats and to develop self-sustaining populations of the threatened taxon (Maunder 1992).

Artificially raised seedlings planted *in situ* have the potential to augment existing populations and increase the number of sites at which a species occurs. In addition, the establishment of artificial populations provides the capacity for experimental manipulation without endangering existing populations. This allows knowledge to be gained on population dynamics, phenology, and ecology that collectively provides valuable strategies for the practical conservation of rare and threatened species.

Since most orchid translocations involve axenically derived seedlings that were grown in highly controlled laboratory environments, the seedlings have not been subject to the rigorous selection processes that occur under natural environmental conditions. Research is required to determine if the genetic diversity of seedlings generated under *in vitro* conditions represents the parental genetic diversity, or whether the lack of ecological selection will result in deleterious genetic drift (Bouzat et al. 2009; Hufford and Mazer 2003).

Case Study 7.1

Peristylus (Platanthera) holochila in Hawaii

Lawrence Zettler
Illinois College, USA

A terrestrial orchid with a history of resisting attempts aimed at its conservation is the Hawaiian bog orchid, *Peristylus holochila* (Hbd.) N. Hallé (syn = *Platanthera holochila* (Figure 7.1). This Federally listed, critically rare, photosynthetic species is endemic to cloud forest bogs at higher altitudes (1,000–2,000 m) of the Hawaiian archipelago (Zettler and Oppenheimer 2012). Fewer than 50 individuals of this species are thought to survive in small populations

Figure 7.1 (a) *Peristylus* (*Platanthera*) *holochila* in flower on Molokai in 2003; (b) three shoot-bearing protocorms of *P. holochila* are shown *in vitro* on a soil/oatmeal-based medium; (c) One of three fungal endophytes (UAMH 10547) recovered from protocorms of *P. holochila* on Molokai from seed packets; (d and e) reintroduction of asymbiotically grown seedlings into Kamakou Preserve, Molokai, in 2011. Students that germinate seeds at Illinois College in their first (freshman) year may be fortunate enough to take part in reintroduction efforts just before they graduate using this slow, cumbersome method (photos: Lawrence Zettler).

scattered on three islands (Kauai, Maui, Molokai). The lone remaining plant on Kauai has persisted for decades as a source of seed for recovery efforts, but there is now concern that it may be in decline as evidenced by the lack of seed production in the past two years. To raise population numbers above critical levels, specialists and orchid growers alike have attempted to cultivate this orchid from seed *in vitro,* dating back to 1977 (Zettler and Perlman 2012).

Initial In Vitro *Germination Trials*—Seeds were sown on asymbiotic media and exposed to light, but embryo development was not successful, possibly because these techniques were more applicable to tropical epiphytes and not terrestrial species like *P. holochila*. In 2002, the focus shifted to inoculating seeds with endophytic fungi and incubation in darkness, paralleling successful studies in Australia, Europe, and North America (e.g., Anderson 1991; Clements et al. 1986; Dixon 1987). This was carried out with Illinois College undergraduate students working alongside Hawaiian specialists Steve Perlman (National Tropical Botanical Garden) and Hank Oppenheimer (Plant Extinction Prevention Program). In the pilot study, seeds of *P. holochila* were sown on an oat-based medium and inoculated with a mycorrhizal fungus from Florida (*Epulorhiza repens* 266), chosen for its ability to facilitate seed germination for a wide range of North American taxa. Protocorms were produced *in vitro* (Zettler et al. 2005; Zettler et al. 2011), demonstrating that the symbiotic technique did indeed have merit for *P. holochila* (Figure 7.1b).

Acquiring Native Fungi (Seed Baiting)—Efforts then shifted to acquiring endophytes from Hawaii, targeting those that the orchid might utilize for germination and development *in situ*. Seed packets (Rasmussen and Whigham 1993, see Chapter 5) placed in soil at the Molokai population yielded protocorms in about one year after burial. Three mycorrhizal candidates were isolated from these protocorms and deposited at the University of Alberta Microfungus Herbarium (UAMH) in Canada as 10546, 10547 (Figure 7.1c), and 19548. GenBank matches for all sequences revealed that these isolates were unnamed *Tulasnella* species (<95% match), raising the possibility that the *Tulasnella* strains in Hawaii may be endemic to the archipelago. None of these fungi, however, were effective in facilitating germination of *P. holochila* seed *in vitro* (Zettler et al. 2011), bringing the project to a crossroad: should the *Epulorhiza repens* strain (266) be included in *P. holochila* recovery plans, given that this fungus originated in Florida?

Reintroduction—The overwhelming consensus of those polled at IOCC2 and by specialists in Hawaii, was to take a more conservative approach and not use a fungus outside of the range of *P. holochila*. Thus, efforts were redirected to using asymbiotic media until effective endophytes from Hawaii could be isolated and tested. Use of medium P723 (PhytoTechnology Labs) resulted in the successful production of 85 leaf-bearing seedlings that were reintroduced on Kauai, Molokai, and Maui (Olinda) in March 2011, 3.1 years after sowing (Zettler and Perlman 2012; Zettler et al. 2011). An additional 56 seedlings were introduced to the archipelago on two subsequent trips the following year; one of these initiated anthesis *in vitro* just prior to deflasking, 59 months after sowing (Zettler and David 2014). Of the 141 total seedlings, two initiated anthesis at Olinda (Maui). Despite this apparent success, asymbiotic techniques remain a slow and cumbersome process compared to the symbiotic techniques applied previously (Figure 7.1d, e).

Future Work—In the decade following the initial work, ITS sequencing (see Chapter 9) identified the Florida fungus (266) as *Tulasnella calospora* (99–100% identity, ITS1, ITS4) and efforts are now under way in Hawaii to analyze fungal DNA in soil samples to determine if this fungus occurs there naturally. If this is confirmed, the next step will be to mass-propagate *P. holochila* from seed using the Florida fungus strain.

7.4 ASSISTED MIGRATION

In conservation planning, balancing rigorous scientific research with the need for immediate conservation action often presents conflicting situations. In addition, there is debate over the ethics and approaches in proactive translocation (assisted migration) of species into a new habitat as a conservation measure.

Migration, in the context of plants, refers to movement beyond the historic range of a plant's distribution, as individuals from populations throughout its range establish at higher latitudes and altitudes in response to movement of their climate optimum (Kirkpatrick and Barton 1997). The process of migration requires both dispersal and successful colonization. Dispersal of a propagule to a new site beyond the parent plant's population can take place without colonization occurring, but this does not constitute a successful migration event.

Often, criticism of assisted migration programs relates to the potential for a migrated species to become a weed (biological invasion). Biological invasions can erode biodiversity and disrupt key ecological processes and interactions. The biological traits that promote endangerment of a species are not simply the opposite of those that favor invasiveness (Jesche and Strayer 2008). Moreover, invasiveness is not necessarily a good predictor of impact; a species that is a poor colonizer in its home range may become invasive when released into new environments.

Given the unique biological and ecological requirements for many terrestrial orchid species, the probability of an orchid becoming invasive is low. However, a number of examples exist (Figure 7.2):

- *Disa bracteata* in southern Australia
- *Microtis media* in mulched gardens in temperate Australia

Figure 7.2 Invasive species (a) *Microtis media*—a native species that has invaded gardens and bushland margins; (b) *Disa bracteata;* and (c) *Zeuxine strateumatica* (photo: Yung-I Lee).

- *Phaius tankervilliae* in Hawaii
- *Zeuxine strateumatica* in Florida and southeast Asia

7.5 FRAMEWORK FOR TRANSLOCATION PROJECTS

Successful orchid translocation requires adherence to a framework of questions and guidelines for consideration prior to and post-translocation (Table 7.1). The framework incorporates decision-making scenarios with pre-translocation assessment, the translocation itself, and post-translocation monitoring. Although each translocation project will vary, and the considerations for each species will be different, this framework provides a solid foundation with appropriate checks and balances.

Table 7.1 Framework for terrestrial orchid translocations: guidelines and considerations

Stage of Translocation	Details	Important Considerations
Is translocation necessary?	Translocation programs require a long-term commitment with the resources required for such programs into the medium, to long term. When investing conservation resources, it is important that the need for translocation is evaluated thoroughly prior to undertaking the program.	• Alternative options attempted or considered • Understand and control threatening processes and risks (see Chapter 2) • Success or failure of previous translocation programs • Costs and resources required for implementing the program • Impacts of the program on neighboring species and the environment
Pre-translocation assessment	The key to effectively and efficiently protect and manage threatened orchid populations is a strong, scientifically based understanding of the relationship between the species and its environment. Once a decision has been made to undertake a translocation program, further information is required to provide an understanding of the biological, ecological, and environmental logistics associated with working on the target species. Given the wide variety of growth forms and mycorrhizal associations in terrestrial orchids, translocations often need to be species specific. Nevertheless, a number of basic factors are common to many translocation programs.	• Pre-translocation monitoring, surveys of orchid distribution, habitat identification (see Chapter 2) • Genetics: taxonomic status, diversity within and between populations/species, propagules (see Chapter 9) • Reproductive biology and pollinator limitations (see Chapter 8) • Mycorrhizal ecology (see Chapter 3), specificity (see Chapter 4), availability and microsite/dispersal limitations, seed baiting (see Chapter 5) • Selecting propagules such as seed, where and how to source the propagules, sampling strategy (see Chapter 4) • Recipient sites for translocation/ reintroduction, suitability, tenure, threats, land use values

Table 7.1 Continued

Stage of Translocation	Details	Important Considerations
The translocation	The translocation itself will vary significantly from one species to another with respect to the considerations detailed above. It is essential that local conservation authorities are consulted, as often a very specific set of criteria has to be met before translocation can occur.	• Translocation proposal submitted to local conservation/management authorities for assessment • Site preparation and amelioration if required • Pilot studies and experimental design as part of a full translocation • Determining number of plants, layout, timing, planting procedures, identification, record keeping • Post-planting care such as watering, mulching, fencing, weed and herbivore control
Post-translocation monitoring, evaluation, and management	Translocated orchids and their habitat require significant ongoing care, management, and monitoring following translocation. Given the resources required to prepare and carry out a translocation program, often post-translocation requirements are forgotten or unable to be followed through. It is vital that the translocation process is documented, including post-translocation results, management, successes and failures, so that an effective post-translocation evaluation can be carried out.	• Preparation and undertaking of monitoring of survival and growth in multiple years • Biological and ecological factors such as survival, growth (height and width), flowering, pollination, recruitment • Maintenance of genetic diversity, ongoing survival of mycorrhizal fungi, presence of suitable and effective pollinators • Post-translocation evaluation and criteria for success, communicating results • Community and stakeholder participation and support, volunteers, education, outreach

7.6 RESTORING HABITATS SUITABLE FOR ORCHIDS

The global decline in orchid habitats means that orchid translocations depend upon an ability to restore a habitat capable of supporting the orchid, it's mycorrhizal agents, and pollinators. Particularly in urban and peri-urban areas, habitat degradation through weeds, altered flows (water, pollinators, nutrients), and changed practices (fire, grazing) mean that there is insufficient suitable habitat or habitat at the scale to enable translocated orchid populations to flourish (self-regenerate in perpetuity).

Ecological restoration (ER), the process of removing or ameliorating degrading processes and reinstating abiotic and biotic elements based on a local native reference ecosystem, is a globally acknowledged and accepted practice. A reference ecosystem is described by the Society of Ecological Restoration as "a model local, native ecosystem that informs the target of the restoration project. The local reference ecosystem provides the specific compositional, structural, and functional ecosystem attributes that enable the restored state to self-organize and ultimately be capable of recovering to the same state as the native model. This model is synthesized from information about past, present, and anticipated future conditions at the site and similar sites in the region, in consultation with stakeholders." ER provides the tools, policy framework, and mechanisms to plan, execute, and assess habitat reinstatement—all processes fundamental to ensuring a net gain in orchid habitat globally.

Guidance on the how, when, what, and who of ER can be found in the newly launched International Standards for the Practice of Ecological Restoration (*The Standards*), published by the peak body for restoration, the Society for Ecological Restoration (see www.ser.org for more details). These Standards provide fundamental tools and concepts for reinstating native, local environments necessary to support long-term orchid conservation. Three underpinning principles (Text Box 7.1) of *The Standards* encourage ER to be:

1. *Effective* in establishing and maintaining an ecosystem's values, species, and processes.
2. *Efficient* in maximizing beneficial outcomes while minimizing costs in time, resources, and effort.
3. *Engaging* with partners and stakeholders promotes participation and enhances a social connection with ecosystems.

An effective tool to gauge how restoration is progressing is the use of the *Recovery Wheel*—developed by *The Standards* group (Figure 7.3). The *Recovery Wheel* is a visualization tool

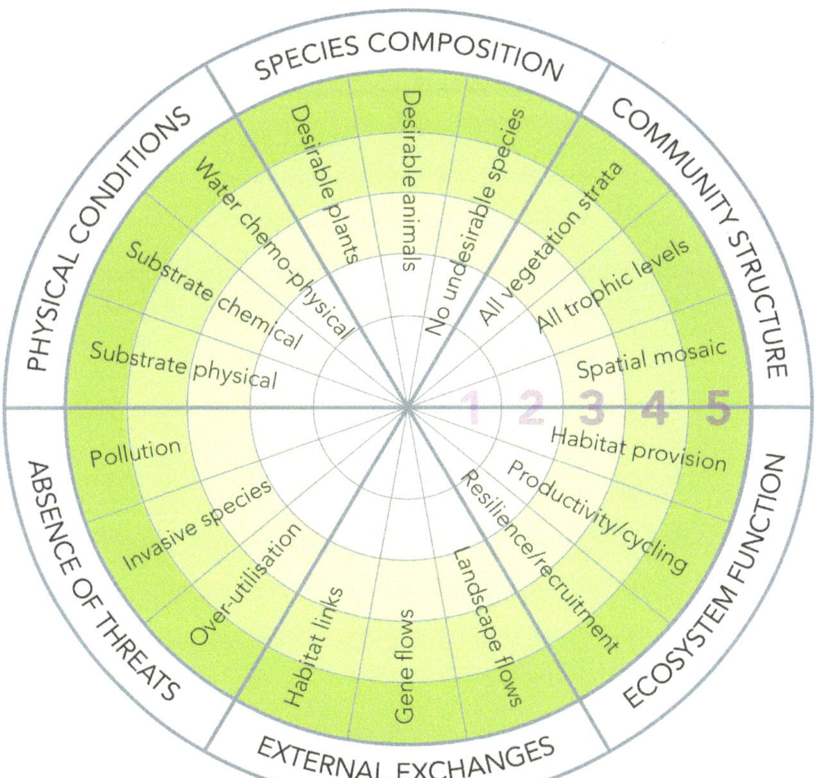

Figure 7.3 The *Recovery Wheel* provides a ready reckoner of how well a recovery/restoration program is progressing toward meeting key goals in restoring an orchid habitat (the *Wheel* is applicable to all ER programs. Go to www.ser.org for further details).

that indicates how restoration is progressing toward the goal of a self-organizing, sustaining habitat. See *The Standards* for details on how to use the *Recovery Wheel* can help guide orchid habitat restoration.

Text Box 7.1

Restoration principles

Six guiding principles provide the framework upon which all orchid restoration programs should be based and are summarized here from *The Standards*:

Principle 1: Ecological restoration is based on identifying and using a local native reference ecosystem as the benchmark for reinstating local species and processes (taking into account environmental change processes such as climate change).

Principle 2: Identification of key attributes to be reinstated in the restoration program based on the local reference ecosystem as a basis for developing longer-term goals and immediate-term objectives.

Principle 3: The most effective means to achieve recovery is to assist natural recovery processes, supplementing where required, when natural recovery potential is impaired.

Principle 4: The restoration effort must seek the highest and best effort toward full recovery of the habitat—this is especially important for orchids that depend upon intricate links to the biotic and abiotic components of an ecosystem.

Principle 5: Successful restoration draws upon all available knowledge, and more important, may need a level of scientific engagement to obtain information critical to ensuring that orchids being translocated into a restored site are self-sustaining; for example, that their pollinators are understood and that the habitat critical for those pollinators is created or restored.

Principle 6: From the outset, that engagement with local communities and stakeholders is undertaken that is reflective of genuine and active dialogue as an underpinning principle of the long-term success of the restoration of an orchid habitat.

Case Study 7.2

Botanic gardens in orchid conservation—integrated orchid conservation at the Atlanta Botanical Garden

Jennifer Cruse-Sanders
Atlanta Botanical Garden, USA

The Atlanta Botanical Garden uses botanical expertise and conservation horticulture to safeguard and propagate rare plant species of the southeastern United States. Applying combined experience in growing rare plants and identification of appropriate microsites for reintroduction or augmentation, the staff at the Garden works with land managers and conservation agencies to restore priority plant species into a variety of habitats. Among the signature plant groups at the Atlanta Botanical Garden is a world-class cosmopolitan orchid collection. Horticulturists at the Garden have developed asymbiotic propagation protocols for orchids, and apply these techniques to propagate native orchid species from across the southeastern U.S. for translocation programs and as an *ex situ* conservation living collection.

North America is home to more than 200 species of orchids, the majority of which are found in the southeastern U.S. One of the nationally listed species, the white fringeless orchid *Platanthera integrilabia,* is a terrestrial orchid that blooms from August to September and is distributed across 50 populations from Kentucky to South Carolina, Tennessee, and Georgia (Figure 7.4a, b). In an effort to conserve populations located in private wetlands within metropolitan Atlanta, Garden staff have partnered with federal, state, and county agencies, utility companies, homeowner associations, volunteers, and students to restore 10 acres of wetland habitat and safeguard plants on protected land. Plants were hand pollinated, the seeds collected, then germinated in the micropropagation laboratory at the Garden, and finally, acclimated in the conservation nursery. Several hundred plants were later translocated into restored sites where the species historically occurred or in adjacent habitats within state-owned parks. One wild population that occurred within a roadside right-of-way was collected to extinction soon after it was discovered in the early 2000s. Fortunately, seeds were sent to the Garden for safekeeping when the population was originally found.

With funding from the National Fish and Wildlife Foundation and the Peachtree Garden Club, these 10-year-old seeds were successfully germinated and plants were reintroduced at the original site. Plants became established and were flowering again at the site within two years of translocation. Ongoing conservation efforts include population monitoring and training local

Figure 7.4 (a) *Platanthera integrilabia* (photo: Brad Wilson); (b) *Platanthera integrilabia* in its natural habitat (Photo: Brad Wilson); (c) *Calopogon tuberosa* (photo: Jeff Talbert); (d) *Pogonia ophioglossiodes* (photo: Jeff Talbert); and (e) *Pogonia ophioglossiodes in situ* (photo: Jeff Talbert).

land management stewards—such as high school volunteers in cooperation with landowners—for long-term management protection of translocated populations. One benefit to the orchid project is that this has acted as a flagship species for promoting more effective management of the wetlands for multiple orchid species. At one restoration site for the white fringeless orchid, more than eight other orchid species were documented in the wetland and surrounding habitat after it was opened up, representing more than 10% of Georgian native orchid diversity.

Within the southeastern U.S., most orchid diversity is located in the state of Florida, however, this diversity is threatened due to habitat conversion and illegal collection. The Atlanta Botanical Garden is partnering with the Florida Park Service to restore 220 acres of wetland communities in Florida State Parks, just outside popular destinations along the Gulf Coast. Garden staff collected seeds from 14 orchid and nonorchid plant species (10 genera) from 16 source populations. Between 2013 and 2016, more than 6,832 plants, 3,317 of them orchids, were reintroduced into the restored areas with an approximately 75% survival rate after at least one year. Restored orchid species included: *Platanthera conspicua*, *Platanthera integra*, *Pogonia ophioglossiodes*, and *Calopogon tuberosa* (Figure 7.4c, d, e). Both partners bring essential elements and skills necessary for the project's success. The Florida Park Service is the landowner and manager. Their team is responsible for carefully clearing woody biomass from degraded wetlands and managing the landscape with prescription fire to restore connectivity between uplands and wetland habitats. The Atlanta Botanical Garden provides project coordination, the restoration plan, project mapping, prioritization, monitoring, and plant propagation to augment what should be a biodiverse herbaceous wetland community.

7.7 METHODS FOR TERRESTRIAL ORCHID TRANSLOCATION

7.7.1 Outplanting Juvenile Herbaceous Terrestrial Orchids

Although translocation requirements vary, usually seed used in germination experiments for translocation will reflect the genetic diversity of the existing natural population. This may have been predetermined through population genetic studies (see Chapter 9) or by using seed sourced from local provenances. In the process of *in vitro* propagation, there is considerable attrition from the number of seeds sown to the number of plants available for suitable translocation. In general, the largest and most vigorous plants grown under these conditions are selected for the program without knowledge of which genotypes are actually being translocated. Genotypes that may be successful under *in vitro* conditions may not be the most successful *in situ*, and further research is needed to determine genetic comparability of artificially generated seedlings.

Early in the growing season (when leaves of two- or three-year-old seedlings emerge), juvenile plants in individual small forestry pots can be transplanted to field sites. If seedlings have been generated symbiotically, mycorrhizal colonization will already be established and will remain with the plant after transplanting, provided habitat conditions are conducive for survival of the mycorrhizal fungus.

The site for translocations should be weed free. Plants should be thoroughly watered before translocation. Supplementary watering may be required depending on the site and climate conditions. It also may be necessary to provide protective cages or fencing, depending on the likelihood of grazing or trampling damage (Figure 7.5).

Figure 7.5 Two types of protective enclosures around translocated plants to prevent grazing

Case Study 7.3

Diuris fragrantissima translocation

Zoe Smith, University of Melbourne, Australia

Background: Diuris fragrantissima is an herbaceous perennial terrestrial orchid endemic to the Victorian Volcanic Plains Grassland in southeastern Australia. With its current distribution restricted to a small section of urban grassland, *D. fragrantissima* is currently listed as endangered and is protected under legislation at both State and Federal levels. Due to the limited success of current management practices in conserving the few remaining extant plants, reintroduction of *ex situ* propagated material was decided on as the preferred conservation approach.

Factors considered before undertaking the study included confirmation of the taxonomic status of *D. fragrantissima* and levels of genetic diversity among source and remnant populations (Smith et al. 2005, 2007). Major threats to the species were identified as loss of habitat and invasion by exotic weed species, the latter of which is controlled by ongoing management of the remnant and recipient sites.

Cultivation of source plants: Adult *D. fragrantissima* plants cultivated *ex situ,* were used as source propagules (seed) for the reintroduction trials. Source plants for the reintroduction were produced by asymbiotically germinating seeds, followed by transfer from an axenic culture to potting media (equal volumes of composted pine bark and perlite). Study plants were tested for the presence of mycorrhizal fungi six months after the transfer to potting media. Symbiotic germination tests and molecular methods were used to confirm the presence of the appropriate mycorrhizal symbiont in the source plant pots (Smith 2006). Plants used in the reintroduction experiments were two to three years old. In all, 600 two- to three-year-old plants and 100 dormant tubers were available for reintroduction.

D. fragrantissima *reintroduction*: A grassland reserve located within the known habitat range of *D. fragrantissima* was selected as an appropriate reintroduction site. The site has vegetation and soil properties similar to known sites of *D. fragrantissima* and is reserved for nature conservation. Source plants were separated into the following three treatments and planted at the recipient site in autumn (April) and spring (September): (1) addition of a mycorrhizal fungus (via millet seed inoculum) to the site soil; (2) addition of a mycorrhizal fungus and soil aeration (tilling with a pick axe); and (3) plants with no additional treatments, serving as a control. Plants received supplementary watering each month and were covered with camouflaged cages to prevent predation. The dormant tubers were planted at the site in the following summer (December) for survival comparison against seedling out-plants. *D. fragrantissima* reemergence was monitored for four years. It was found that reemergence across all treatments decreased at an average of 16.9% per annum. Reintroductions of *D. fragrantissima* in spring were more successful when a fungus was added and the soil aerated, and in general, survival and growth of plants reintroduced in autumn were best when no treatment was applied (>40% survival after four years). It was recommended that for the efficient use of resources, future reintroductions of *D. fragrantissima* be undertaken in autumn. This study highlighted the value of conducting experimental treatments for determining reintroduction success and future management actions (Figure 7.6).

Figure 7.6 (a) *Diuris fragrantissima* in pots and (b) *Diuris fragrantissima* translocated into field sites

7.7.2 Outplanting by Tuber/Rhizome Transfer

Dormant juvenile tubers of terrestrial orchids produced following the methods described in Chapters 4 and 5 can be planted in field sites prior to dormancy break. Disturbance to each site is minimal, as dormant tubers are inserted into the ground using a small cylindrical auger approximating the size of the tubers. This method is most appropriate at transfer locations where compatible mycorrhizal presence has been confirmed, or where tubers are being transferred to sites in close proximity to existing mature plants. As tubers/rhizomes are generally not colonized by mycorrhizal fungus, fungal inoculum needs to be supplemented at sites where mycorrhizal presence has not been confirmed. This can be done in two ways.

1. Three 3 mm cubes of oatmeal agar (OMA) culture media colonized with the mycorrhiza are placed in small holes (1–2 cm deep) adjacent to translocated tubers to act as a source of inoculum.
2. The site can be inoculated using millet seed containing the compatible mycorrhizal fungus. Preparation of the millet seed inoculum is described in Chapter 5.

Again, supplementary watering may be required, and sites may need to be protected against rodents and birds. These may be attracted to the millet seed inoculum and can cause damage to a translocation plot if protection measures are inadequate.

7.7.3 *In Situ* Seed Sowing

An alternative to translocating individual plants is to generate orchid populations through direct *in situ* seed sowing. This method may be most suitable if a laboratory or glasshouse is not available to grow *ex situ* plants. However, this can only be done if large amounts of seed have been collected since the method can be particularly wasteful of seed.

Seed is sown to pre-inoculated field sites at the commencement of the growing season. As described earlier, inoculating sites with millet seed containing compatible mycorrhizal fungi or sowing adjacent to an adult plant is necessary to facilitate germination in many terrestrial taxa.

As orchid seed is dust-like and easily wind-dispersed, care must be taken when sowing *in situ* to ensure it falls in the required location. To assist sowing, seed can be thoroughly mixed with fine-grained clean sand and a pepper shaker used to evenly spread the mix over the soil. A thin layer of finely chopped mulch appropriate to the natural mulch occurring at the site should be applied over the freshly sown seed, and the soil should remain moist at all times. Seedlings that germinate and establish under these conditions may be the best adapted to the conditions, increasing the chances of survival over the dormant period. If too many seeds germinate, individual seedlings may be pricked out and transferred to a nearby suitable location to reduce competition.

Case Study 7.4

Wild orchids from seed spread 'on the wind'

by Dave L. Trudgill
The Steading, Blairgowrie, Scotland

In 1998, we took ownership of a small, 0.2 hactare field in eastern Scotland and began the conversion into a wildflower meadow. The area was sprayed twice with herbicide before being sown with a wild flower mix and a small pond was dug. Four years later the spoil heap from the pond was covered with northern marsh orchids, *Dactylorhiza purpurella*—the orchids were not present prior and, therefore, must have arrived as windblown seed.

A further eight species of orchids (each suited to grassland sites) have since been added to the meadow. All were grown from the seed of local species (within 50 km) that was hand broadcast over the site—downwind. Transplanting living material was avoided to reduce the possibility of introducing disease. The first major success was the flowering of a rare British orchid, the lesser butterfly orchid, *Platanthera bifolia*, in 2006. Two more plants appeared in 2007 and by repeatedly respreading the seed from this species, the number has increased to more than 60.

Seven other orchid species have been successfully introduced to the meadow using hand broadcast seed. These are the common spotted (*D. fuchsii*), heath spotted (*D. maculata*), early marsh (*D. incarnata*), early purple (*Orchis mascula*), greater butterfly (*Platanthera chlorantha*), heath fragrant (*Gymnadenia borialis*), and marsh helleborine (*Epipactis palustris*) orchids. Some of the nine species have increased greatly in number, especially the northern marsh and common spotted orchids, so that there are now more than 2,000 orchids. Others have not done so well, especially the heath spotted orchid, which favors a more acidic soil. Each species sown has produced flowering plants; consequently, the fungi involved in the development of these orchids are either very widespread and/or the associations are not specific.

The meadow was previously part of an arable field and it was essential to decrease the fertility of the site. This was done by mowing the meadow each autumn (after the orchid seed had been collected) and removing all the cuttings as hay. Decreasing the fertility was a slow process and even after nearly 20 years, some parts still produce a reasonable crop of hay, but few orchids. Vigorously growing perennials were also targeted and eliminated and a seed mixture based on native, wild grasses that lacked legumes but included yellow rattle (*Rhinanthus minor*), a semi-parasite of grasses, were chosen. The meadow was fenced and netted to exclude herbivores. At sites where mowing was not possible, grazing (preferably by sheep, as cattle cause more treading damage) during the autumn and winter was required.

The meadow was sown with orchid seed by progressively opening orchid seed capsules while moving around, taking account of the wind speed and direction. The seed heads were collected when *ripe* (i.e., changing color from green to yellow/brown), but before they had started to shed seed. Where only small amounts of seed were available, the orchid seed was mixed with dry, sieved leaf-mold, thereby facilitating a more even spread of the seed over a wider area.

The meadow is special in that it provides a diversity of habitats. It is on a gentle south-facing slope, with mostly a light loamy soil with a pH of 5.5 to 6.0. However, the pond at the bottom of the meadow is fed by water from the Lunan Burn and has a pH of around 8.0. Consequently, the pond edge provides a damp/marshy habitat with a pH ideal for the marsh helleborine and early marsh orchids. Rushes are eliminated by covering some parts with black polythene sheeting during the orchid dormant period. There are also a few trees in the meadow which provide shaded areas favored by the early purple orchid. Removal of the topsoil from part of the meadow and adding large amounts of crushed limestone prior to spreading the seed of calcicolous species is now being investigated.

Orchids have declined greatly in the UK, largely due to changes in land use. More needs to be done to protect, conserve, and increase their abundance and range. In a time of climate change, this should be done in a way that increases understanding of their requirements and limitations and should be appropriately recorded as there is still much to learn about their longevity, habitat requirements, and management (Figure 7.7). However, our experience shows that with careful planning and the removal of ecological impediments (high soil fertility; overgrowth), it is possible to build a highly effective and productive orchid meadow that has the potential to add to the conservation of British orchids.

Figure 7.7 (a) *Platanthera bifolia* (lesser butterfly orchid) in marsh; (b) *Epipactis palustris* (marsh helleborine) growing on the side of the pond; and (c) Summer view over the pond showing grass and herb growth (photos: Dave Trudgill).

7.7.4 Transplanting Mature Wild Plants

Transplanting (salvage) from a wild site is the conservation action of last resort with habitat preservation and other protective conservation actions taking precedence. If transplanting (salvage) needs to occur, individuals stand a better chance of survival under managed *ex situ* conditions in a glasshouse or greenhouse, prior to transfer to a new site. This also provides an ongoing source of propagules and seed orcharding opportunities. There is relatively little known about orchid translocation, especially in relation to mycorrhizal interactions, therefore the methodology is designed around optimizing the survival of plants.

For herbaceous terrestrial orchids, a method for transplanting mature plants is extracting individual tubers while they are dormant and transferring to habitat-matched locations containing suitable mycorrhizal fungi (see sections on artificial inoculation and mycorrhizal baiting). Individuals for transplanting need to be marked by GPS and small pegs to assist in relocating the plants during their dormancy period. When plants are dormant, the topsoil around the dormant plant is removed and collected. Then, with care, tubers/rhizomes can be carefully excavated and stored in a dry, dark location with padding to prevent bruising. It is also possible to extract a divot containing the orchid tuber if you are confident of the depth and position of the dormant organ of the orchid.

At the recipient site, tubers are transplanted to the same depth from which they were excavated for each individual. Following tuber placement, soil and organic matter collected from the source site should be replaced to maintain preexisting soil profiles. If required for long-term monitoring, locations should be permanently marked with a metal stake at a set distance and direction from the plant to assist in relocation of the plants.

Alternatively, the plants can be carefully excavated and transferred to pots for storage under nursery conditions until they can be transplanted at the recipient site. This method can be used either while the plant is dormant or as the plant reemerges early in its growing season. With all transfers involving habitat soils, strict phytohygiene standards are

Figure 7.8 (a) Excavation of dormant orchid tubers. Dried and dead tuber remains (*husks*) in *Caladenia huegelii* indicate this orchid is approximately 20 years old and (b) difficulties with excavation include tree roots.

necessary, and extreme care should be taken to ensure that the donor site is free of pests and disease (Figure 7.8).

7.8 PRE- AND POST-TRANSLOCATION SURVEYING AND MONITORING

Monitoring techniques for terrestrial orchid populations have been described previously in Chapter 2. The data to be collected for pre- and post-translocation monitoring should be defined in reference to a predetermined set of objectives or questions, aimed at filling knowledge gaps about a species or population and its management. For example, data collection should take place before translocation to determine the extent of a species habitat range, how many plants are in a population, which plants are reemerging, and what are the ecological characteristics of successful plants within the population. These data are used to derive demographic patterns and trends, and are useful for comparison of performance of translocated individuals against benchmark populations in natural habitats. In addition, through the use of scientifically rigorous design, it is possible to experimentally test the effectiveness of a range of management actions aimed at addressing a particular threat.

The year-to-year success (Table 7.2) of the translocation of orchids can be evaluated by plant reemergence in the growing season and comparison of growth parameters such as leaf length (Figure 7.9) and width, and the reproductive potential between transplants and individuals. By taking similar measurements on a yearly basis from individuals in neighboring benchmark populations, a statistical comparison can be made on translocation growth and development. The underground stem or roots of translocated plants can be sampled for mycorrhizal reinfection. This method can track the persistence of fungal inoculum and is suitable for all types of translocations. Typically, for tuberous herbaceous perennial orchids, a plant can survive on tuber reserves for up to three years without an effective mycorrhizal partner—but considerably less for juvenile plants. Since this method is destructive, additional individuals may need to be incorporated into the program to facilitate monitoring.

Table 7.2 Criteria for assessing the long-term success of orchard translocations (based on Reiter et al. 2016)

Years Post-Translocation	Success Criteria
1–3	Sufficient plants survive to be considered a viable population (i.e., for out-breeding long-lived species, 50–200 individuals (Neiland and Wilcock 1998)). Representing a genetic spread of the source population or species. Rate of emergence stable after initial deaths related to translocation shock (post 1st years emergence), emergence comparable or above wild population.
3–7	Plants established, flowering, and evidence of fruit set. Emergence rates, flowering and fruit set not significantly lower than benchmark population.
8–14	Recruitment seen in the recipient site. Emergence rates stable, flowering and fruit set not significantly lower than benchmark population.
15	A self-sustaining population is established. Total number of new recruits equals or exceeds mortality. Population growth rate or equal to 1. Area occupied is stable or increasing.

Figure 7.9 Post-translocation management: collecting performance data of transplanted individuals

If mycorrhizal fungi are not detected at the translocation site, remedial action, such as the supplemental addition of millet seed inoculum, may be required.

Other parameters for tracking post-translocation performance include recording flower production and seed set to determine the reproductive potential of translocated plants. When this potential drops below benchmark data, as determined by appropriate statistical analysis (e.g., analysis of variance), remediation in the form of pollinator encouragement through the planting of food plants may be necessary. Hand pollination can overcome declines in seed set, however, care is required to ensure appropriate levels of out-crossing occur.

Case Study 7.5

Reintroduction of Cypripedium macranthos *in the mountains of Beijing*

Zhang Yu
National Engineering Research Centre for Floriculture, Beijing Forestry University

Cypripedium macranthos (Figure 7.10a, b) is regarded as one of the most beautiful orchids in north China and is extremely difficult to find due to its rarity. In the mountainous area of Beijing, there are a number of small remnant populations where conservation efforts can be focused. Following IUCN standards, *C. macranthos* had a previous ranking of Near Threatened (NT), however, this was upgraded to Vulnerable (VU) a2c in 2004. It is now recommended that the species be further upgraded to the Endangered (EN) category mostly due to its severe extinction risk as long periods of over-collection have destroyed most of the remaining wild populations.

Figure 7.10 (a) *Cypripedium macranthos*; (b) *Cypripedium guttatum*; (c–h) the process of seed germination from seed to seedling in *Cypripedium macranthos*; and (i–k) seedlings in different compost treatments (photos: Zhang Yu).

Asymbiotic seed germination protocols were originally used to successfully germinate seedlings, however, large losses at the deflasking stage meant that the method was not reliable or suitable for this species (Figure 7.10c–h). Recently, with the use of mycorrhizal inoculation of germination media, successful germination has been obtained. Mycorrhizal fungi were isolated from root material and soil. Identification using DNA sequencing of the nuclear ITS gene region has revealed the suitability of some mycorrhizal isolates for seed germination.

After deflasking, young seedlings are sent to a small *Cypripedium* nursery in the mountain area in north Beijing, which provides a halfway house between the laboratory and field sites (Figure 7.10i–k). After two years of growing outside, some seedlings were transplanted to their original habitat early in 2015. This study will take a number of years to determine if the transplanted seedlings are long-term survivors capable of becoming a resilient and self-sustaining population.

8

ORCHID POLLINATION

8.1 BACKGROUND

Orchids are renowned for their spectacular floral diversity that represent elaborate and elegant adaptations for pollination. The study of orchid pollination, particularly from an evolutionary perspective, is advanced in comparison with other biological and ecological aspects of this family.

Darwin (1862) delved into the relationships between orchid floral diversity and complexity of pollination strategies, considering their adaptations for pollination as examples of evolution through natural selection. Since then, a history of scientific enquiry into orchid-pollinator interactions has demonstrated the extraordinary degree to which orchids have refined pollinator attraction. Pollination research requires a multifaceted approach incorporating developmental biology, community ecology, animal behavior, floral evolution, genetics, and ecosystem studies.

Pollination research developed from two historical approaches. The emphasis of the first approach was detailed observation of floral mechanisms and the natural history of the ecological relationships between plants and pollinators, while the second focused on evolutionary processes that might influence and be affected by pollination. Today, many orchid pollination studies are undertaken within a mechanistic framework that investigates both ecological and evolutionary questions (see reviews by Gaskett 2011; Jersakova et al. 2006; Tremblay et al. 2005).

Little research has been undertaken into orchid pollination in a conservation context. While evolutionary questions are generally the drivers of pollination research, answers to conservation questions often require a different approach. For example, the asymmetry between orchids and their pollinators where orchids are far more reliant on their pollinators than the pollinators are on the orchid, requires careful conservation planning to ensure pollinator requirements such as food plants, nest sites, larval host species, or larval host plants are sustained in conservation and translocation programs. Additionally, identifying orchid pollinators and their habitat requirements in the context of habitat loss and degradation is crucial to the restoration of habitats and orchid establishment at these sites. This chapter covers practical methods applicable to orchid pollination research from pollinator baiting and capture techniques to an introduction to pollinator pheromones and their use in pollinator baiting studies.

8.2 POLLINATOR LIMITATION

Much of the diversification of the flowering plants is attributed to the development of animal-mediated pollination (Galen 1999; Grant 1994). While multiple reproductive barriers (pre-zygotic and post-zygotic) typically isolate plant species, a unique feature of animal-pollinated plants is the potential for pollinator-mediated isolation to drive diversification (Darwin 1862; Peakall 2007; Rieseberg and Willis 2007). Yet, despite growing evidence for pollinator-driven speciation, we still have much to learn about the process, in the context of both speciation and conservation.

Given that most orchids use animal vectors to promote outbreeding, pollinator and resource limitation are major issues in conservation planning. Tremblay (2005) showed that although resource constraints (nutrients, light, moisture availability) can limit orchid abundance and distribution (Montalvo and Ackerman 1987), pollinator limitation appears to be the dominant factor limiting reproductive success. Studies in which flowers are subject to supplementary cross-pollination showed an increase in fruit set compared with natural populations in 56 out of 57 orchid species (Tremblay et al. 2005). Pollination limitation or failure in plants can be attributed to the absence of pollinators, environmental loss of pollen between the pollen donor and recipient, poor quality and insufficient pollen quantity, or the presence of heterospecific pollen (pollen of other non-orchid plants) on recipient plants (Wilcock and Neiland 2002), while resource limitation may be a reflection of the fitness costs associated with sexual reproduction in the previous season.

Many mechanisms are found in orchids to promote pollination success under conditions of pollinator limitation. Some of these include extreme pre-pollination floral longevity (as found in the month(s)-long flowering in species of *Paphiopedilum*), prolonged viability of excised pollen, extended stigma receptivity, and delayed development of ovules. However, human impacts through habitat degradation or fragmentation and over-collecting may interrupt reproductive performance by reducing pollen movement with a subsequent decrease in fruit set and seedling recruitment. Thus, an understanding of limitations on reproductive success in orchids, and the anthropogenic or environmental impacts on this system, is important in conservation planning and management.

8.2.1 Testing for Pollination Limitation

The study of pollination limitation in plants is undertaken generally in the context of evolutionary processes (Tremblay et al. 2005). However, studies of pollination limitation from an evolutionary viewpoint will also apply to the conservation of rare species. Once the cause of pollination limitation or failure is known, management strategies may be implemented to enhance the likelihood that fruit set will occur.

8.2.1.1 What Is the Frequency of Pollination?

With the exception of orchids that are apomictic and self-pollinate, successful sexual reproduction requires a vector to transfer pollinia. Widespread pollinator limitation in orchids is often attributed to insufficient pollinator visits capable of delivery of compatible pollen (Ackerman and Zimmerman 1994). When pollinator abundance is low, or if the orchid is not effective at attracting pollinators, some flowers or individual plants within a population may never be visited. In many species, the ratio of pollinia removals to pollination events can be very low.

To test for pollen limitation by the frequency of pollination visitations, a set of parameters are quantified in a range of populations at set time intervals, usually over a number of seasons (to understand inherent variability in pollination between years). These parameters include pollinator efficiency (pollen removal and deposition rates) and reproductive success (fruit set). The first test is to determine if flowers are capable of elevated pollination levels. This is undertaken by hand pollination (Figure 8.1) to determine if fruit set is elevated in response to an increase in levels of pollen transfer. Reproductive success can then be determined by comparing the success of fruit set following hand pollination versus open (natural) pollination.

8.2.1.2 Determining the Quantity of Pollen Reaching the Stigmas and Effecting Fertilization

Even with frequent visits by a pollinator, the quantity of pollen being transferred may be insufficient to promote fruit set. A pollinium contains many thousands to tens of thousands of pollen grains, however, the quantity of pollen deposited may be insufficient to fertilize all available ovules. Increasing the quantity of pollen deposited on a stigma has been found to increase fruit set in natural populations of orchids (Gregg 1991).

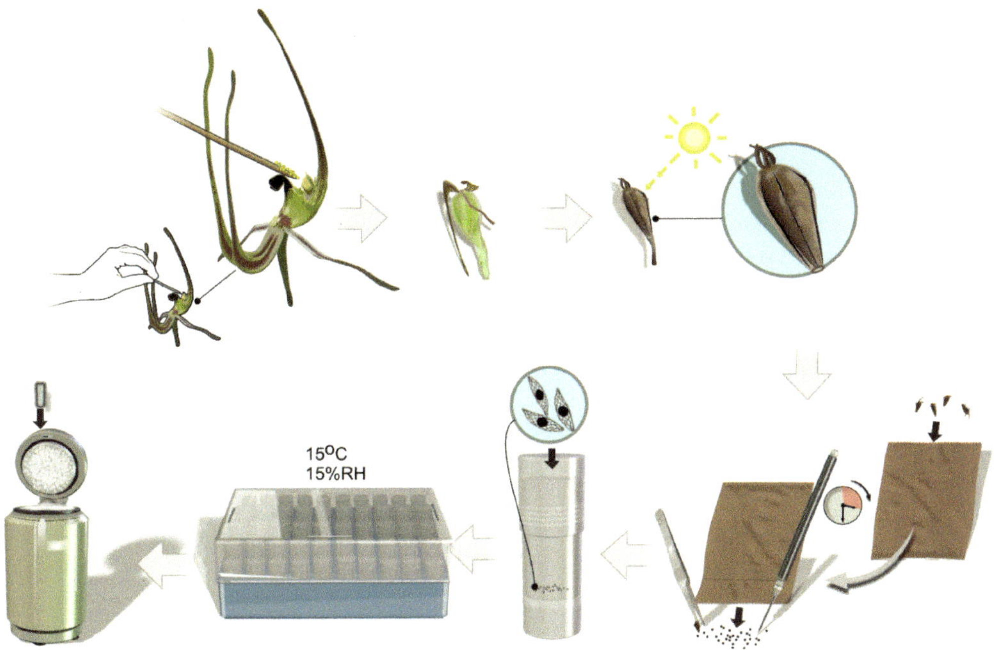

Figure 8.1 Hand pollination protocol that leads to cryopreservation of seed in cryo-vials in the vapor phase of a liquid nitrogen container

Text Box 8.1

Testing for pollen quality: pollen to ovule ratio

One of the ways to test for pollination limitation by pollen quality is to determine the pollen to ovule ratio (P/O). This ratio reflects the likelihood of sufficient pollen grains reaching the stigma to sustain fruit set (Cruden 1977). The more efficient the transfer of pollen, the lower the P/O should be. Although P/O ratios have been considered a useful indicator of a plant's reproductive capability and breeding system, in some orchids and other plants with pollen aggregated into pollinia, P/O ratios may not always follow Cruden's principle (e.g., Wyatt and Broyles 1994).

To determine the P/O ratio, all the pollinia from at least ten flowers from each species, selected from a different individual, need to be collected and soaked in 10% potassium hydroxide (KOH) for 5 hours before macerating. Pollen grains are sub-sampled in 200 ml of Alexander's stain (Alexander 1980) and counted using a haemocytometer. Capsules in the early ripening stage are used to determine ovule numbers. This is done by extracting ovules from one carpel in 500 ml of water and counting under a stereomicroscope. P/O ratios can then be determined by dividing the estimate of the number of pollen grains per flower by the number of ovules per flower. See Cruden (1977) for further information on this technique. A field proxy to the P/O ratio can be determined by estimating the efficiency of fruit set. This is calculated as the ratio of flowers that set fruit to the number of flowers where pollen had been removed through open pollination in a population.

8.2.1.3 What Is the Quality and Diversity of the Pollen Reaching the Stigma?

Following pollen deposition on the stigmatic surface, pollen tubes must grow, followed by fertilization of ovules leading to seed formation. The source of pollen (whether arriving from the same plant, i.e., self-compatible, or a close relative, i.e., cross-pollination, or from an unrelated donor) can influence fruit and seed set and even the vigor of offspring (Charlesworth and Charlesworth 1987). There are a variety of causes whereby pollen quality may influence pollination limitation and therefore seed production rates. These include: self-incompatibility, competition from foreign pollen, male sterility, pollen quality variation, and the presence of recessive lethal genes.

To compare the impact of pollen quality on fruit set, self-compatibility, cross-pollination, and natural pollination can be compared and analyzed. To test for self-compatibility, pollinia must be removed from the flowers and then the flowers bagged until the stigma becomes receptive (in orchids this is often at the same time as pollen removal). When flowers are considered receptive (when the stigmatic surface is sticky and moist) they can be hand-pollinated with pollen from another flower on the same inflorescence and bagged again until wilting (usually a matter of days). For cross-pollination, the same treatment applies, but flowers are pollinated with pollen from another plant. For natural pollination, open pollinated flower buds should be tagged and allowed to develop to seed maturation (seed-proof bags may be attached to maturing seed capsules to maximize seed recovery).

Poor pollen quality can result from male sterility, variation in pollen age, and presence of recessive alleles (Figure 8.2). Though still resulting in some fruit set, these factors may

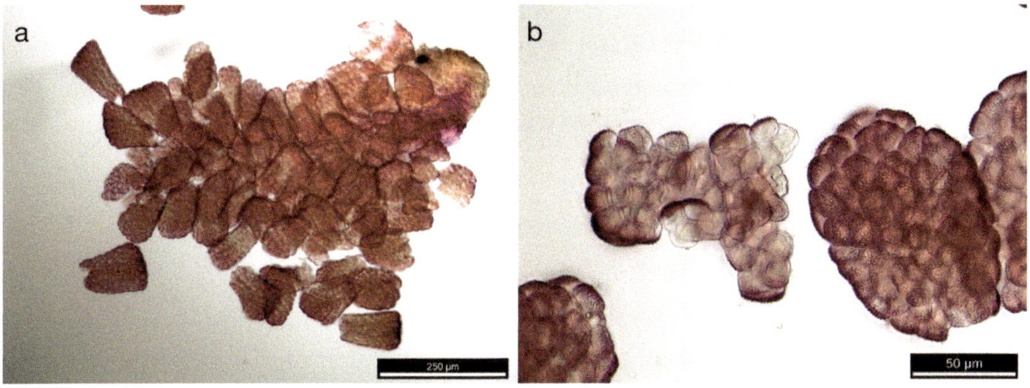

Figure 8.2 Orchid pollen (*Leporella fimbriata*) stained with Calberia's fluid to enhance contrast at: (a) 100× magnification and (b) 400× magnification (photos: Royal Botanic Gardens, Melbourne).

influence the mass of mature fruits, seed mass, and seed quantity (Tremblay 1994). The presence of heterospecific pollen has been shown to reduce seed set by 10% due to fertilization failure. Neiland and Wilcock (1999) found that six out of every seven species tested had pollen contamination in over 85% of stigmas, which could include pollen from plants other than orchids. Contamination can come from true pollinators that have visited non-orchids, wind-dispersed pollen, and from flower visitors other than true pollinators. Although this is well-known, its long-term effect on orchid population viability is unclear.

Pollen diversity can be increased on the stigma where pollen from multiple parents is deposited on the stigma. The incidence of this multi-parent deposition is probably infrequent (though more likely where pollinators accumulate pollinia from different flowers) as flowers generally senesce soon after pollination (Nilsson 1981). For example, multi-paternal pollination resulted in higher germination rates although embryo sizes were generally smaller in artificially pollinated *Cypripedium* (Tremblay 1994).

8.2.2 Resource Limitation

While many species are generally pollinator limited for a flowering period or season, sexual reproduction through pollination is likely to be resource limited over time (Roberts 2003). Resource limitation in orchid populations occurs when supplementary pollination fails to increase fruit production without an augmentation of resources (Tremblay et al. 2005). Supplemental pollination is achieved by using all the flowers on an individual. This reveals whether or not the plant sequesters resources away from less intensely pollinated flowers in favor of those artificially pollinated.

Seed production comes at an energetic and nutrient cost to the parent plant with an increase in fruit production in one year, potentially negatively affecting growth and reproduction in subsequent years. Calvo and Horvitz (1990) suggested a trade off between elevated fecundity gained through increased fruit production and subsequent negative impacts on parental plant vigor. Evidence for (Ackerman 1989; Ackerman and Montalvo 1990) and against (e.g., Calvo and Horvitz 1990; Gregg 1989) this argument has been demonstrated in a number of orchid species with no clear pattern emerging as to which orchid species are likely to be adversely affected through artificially elevated pollination and seed set.

Environmental factors have a significant impact on resource availability in orchids both within a season and also over the life span of the orchid. Nutrient availability (Mattila and Kuitunen 2000), water stress (Mattila 2000), and light intensity (Zimmerman 1980) have been shown to cause resource limitation responses in orchid seed production. In some terrestrial species, episodic events, such as fire, may also have consequences on orchid pollination through increasing light availability and triggering the release of phytoactive compounds such as ethylene that promotes flowering in the first season postfire (Dixon and Barrett 2003). Identifying factors that influence seed output can require intensive multi-seasonal research with such information also indicating how orchids will respond to future environmental change and predicate management strategies (Roberts 2003).

8.3 POLLINATION STRATEGIES

Most orchids are pollinated through the offering of a reward, however, a third of the family are rewardless; instead utilizing a variety of deception strategies to attract their pollinators. Up to 10% of orchids self-pollinate though this may be higher with further careful research (Figure 8.3).

Pollination syndromes can also play a role in orchid rarity. For example, in a study comparing the occurrence and distribution of nectarless and nectar-rewarding orchids in Britain, Neiland and Wilcox (1998) found that a similar proportion of nectar and nectar-less species had widespread or restricted distributions, but only one of the 11 rare orchids produced nectar. This suggests that orchid rarity was more prevalent in species adopting a more specialized deceptive pollination strategy. Phillips et al. (2011), found that pollination strategy showed no significant relationship with the mean abundance or mean distributional extent of orchids in the orchid-rich global biodiversity hotspot of southwest

Figure 8.3 Four pollination strategies represented in terrestrial orchids in the global biodiversity hotspot of southwest Australia (a) self-pollination—*Microtis media*; (b) food rewarding—*Prasophyllum fimbria*; (c) food deceptive—*Caladenia longicauda* subsp *eminens*; and (d) sexually deceptive—*Drakaea livida*.

Australia, but that a higher incidence of rarity was found in the most specialized deception pollination strategies (sexual deception) with almost no rare taxa providing a food reward. Although these studies represent orchids from just two bioregions, there is a trend for more specialized orchids to be the rarest (Swarts and Dixon 2009). This is particularly important in the context of habitat loss and degradation, climate, and environmental change, where small changes to one aspect of the ecological system can have ripple impacts on orchid survival. As methods for determining pollinators for food rewarding species are similar to those for food deceptive species, refer to Section 8.3.3.1 for methods (see Figure 8.4).

8.3.1 Pollination by Reward (Food Advertisement)

For successful reproduction to occur, flowering plants generally utilize the services of pollinators through the offering of food such as nectar, oils, and pollen (Simpson and Neff 1983). A variety of animals from insects to birds learn to associate floral signals such as flower morphology, color, fragrances, and function with the rewards they receive from visitation. For orchids this is true for the majority of the family with over 15,000 species thought to offer some sort of food reward to their pollinators.

The fitness advantage associated with offering a reward relates to imprinting on the pollinator that a reward is only associated with one flower type. The pollinator restricts visits to this one flower type, even when other rewarding floral types are available. This type of pollinator behavior limits the transfer of the pollen of other orchid and non-orchid pollen on the orchid stigmas (Jones 2001). While flower constancy is not expected to have played a role in orchid speciation, a result of this behavior is likely to be stabilized selection on floral signals where nonuniform phenotypes are ignored by the pollinators (Salzmann et al. 2007).

Tremblay et al. (2005), identified three types of floral rewards in orchids. The first and most abundant is nutritional, which can be in the form of nectar, oils, pollen, or pseuodopollen that are consumed by the pollinator. The second is specific to male euglossine bees that are attracted by floral fragrances that they collect for mate attraction. Third, some species such as *Maxillaria* produce waxes and resins for collection by wasps and bees, with these materials thought to be used for nest construction.

Figure 8.4 Food deceptive orchids (a) *Anacamptis morio*; (b) *Barlia robertiana*; (Photos: Jacopo Calevo) (c) *Neotinea ustulate*; and (d) *Orchis militaris* (photos: Julienne Schiebold).

8.3.2 Pollination by Deception

The ability to attract pollinators without offering a reward has evolved independently in many angiosperm lineages, but usually in just a few species per family (Renner et al. 2006). Evidence from phylogenetic studies points to deceit pollination as an ancestral trait in orchids, or that there have been repeated divergences in the orchids (Cozzolino and Widmer 2005b).

The mechanisms of deception in orchids include generalized food deception, food-deceptive floral mimicry, brood-site imitation, shelter imitation, pseudo-antagonism, rendezvous attraction, and sexual deception (Jersakova et al. 2006). Of these mechanisms, generalized food deception is the most common (38 genera) followed by sexual deception (18 genera).

Text Box 8.2

Deception and plant fitness

Three principal hypotheses have been proposed as to how deception could increase fitness in plants:

1. Resources are reallocated from reward production to flowering and seed production (Jersakova et al. 2006).
2. Higher levels of cross-pollination due to pollinators visiting fewer flowers on nonrewarding plants results in higher levels of outcrossing and more efficient pollen export (Jersakova et al. 2006).
3. Deceptive pollination leads to greater pollen removal as the pollinator maneuvers in the flower, seeking the phantom reward (Smithson and Gigord 2001).

In most plants, the established pattern for *rewardlessness* is low pollinator visitation rates and consequently, low levels of fruit production. In addition, and more specifically for orchids, both food and sexual deceit are associated with low fruit set (fruiting failure of 66.5% and 29.1% respectively in deceptive orchid species compared to rewarding species (Tremblay et al. 2005)) and populations appear to be pollinator-limited (Neiland and Wilcock 1998).

Why is pollination by deception so over-represented in the orchid family? Most explanations are associated with the inherent behavior of deceived pollinators. Insects often spend a longer time on rewarding plants and tend to visit more flowers from the same plant and neighboring plants in a patch. For deceptive orchids, the pollinator often immediately departs the patch upon discovering the lack of reward (Ayasse et al. 2000; Johnson 2000; Peakall and Beattie 1996). This effect is even more evident in sexually-deceptive orchids (Peakall and Beattie 1996; Peakall and Schiestl 2004). This behavior increases the chances of outcrossing and long-distance pollen flow as pollinators are more likely to transfer pollen to an orchid in a different patch.

To repeatedly attract pollinators that may learn to avoid nonrewarding flowers, many orchids employ the strategy of mimicry, which imitate signals attractive to pollinators. There are two types of mimicry recognized among deceptive orchids:

- Batesian mimicry that consists of a *mimic* imitating signals of a *model*, and an *operator* that responds. Here, the reproductive success of a mimic depends on its frequency relative to the model, and the inability of the operator to discriminate between the two. For example, the South African orchid, *Disa pulchra,* has nonrewarding flowers that are remarkably similar in morphology and spectral reflectance to the flowers of a sympatric nectar-producing iris, *Watsonia lepida* (Johnson 2000). Both *D. pulchra* and *W. lepida* are pollinated by the long-tongued fly, *Philoliche aethiopica* (Tabanidae), which were unable to discriminate between the two species in choice experiments (Johnson 2000).
- Generalized mimicry, where plants copy general floral signals such as bright colors and floral scent. Here, plants often bloom gregariously and rather than resemble any of the co-blooming species, produce large, showy floral displays (Figure 8.5). Although generalized mimicry is assumed to be common among food deceptive plants, Batesian mimicry has been demonstrated only rarely among food deceptive species, but is well established in sexually deceptive (SD) orchids.

8.3.3 Food Deception

The discovery of pollination by deceit in 1793 by Sprengel involved orchids that possess nectarless floral spurs (Sprengel 1793). Although these findings were initially met with disbelief by Darwin, he eventually established that some orchids successfully attract pollinators without providing a reward (Darwin 1862).

Generalized food deception is the attraction of pollinators by signaling the presence of a food reward, such as nectar or pollen, without ever providing the reward. Orchid species are considered food deceptive on the basis of the absence of detectable nectar and the failure to attract sexually deceived pollinators. These orchid species may resemble a specific co-occurring rewarding species (Johnson 2000). More often, food-deceivers rely

Figure 8.5 Generalized floral mimicry. *Diuris magnifica* and a co-occurring pea flower *Bossiaea eriocarpa* from southwest Australia.

on a general resemblance to food-rewarding species with large bright flowers that exploit the innate searching preferences of pollinators (Nilsson 1992; Schiestl 2005). Food deception has evolved multiple times across angiosperm lineages, but is typically restricted to a few species per family. By contrast, one-third of orchid species are thought to employ food deception.

8.3.3.1 *Methods for Assessing Food Deception in Orchids*

A variety of invertebrates are employed in the pollination of food deceptive orchids. Bees are the most frequently encountered, but butterflies, moths, flies, and beetles are also exploited by orchids.

The attraction of insects to food-advertising orchids and subsequently their capture is often serendipitous; however, dedicated observation has led to major advances (see Kuiter 2013). Molecular approaches can also assist in assigning pollinators to orchids. For example, studies in *Diuris* (Indsto et al. 2006; Indsto et al. 2007) using amplified fragment length polymorphism (AFLP) fingerprinting was able to accurately identify which pollinators were visiting flowers. A study of nonrewarding *Aanacamptis morio* (Johnson et al. 2004) examined pollinator visitation and pollen transfer by queen bumblebees with the addition of nectar to spurs of *A. morio*, significantly increasing the number of flowers probed by bumblebees.

Visual signals are thought to be of primary importance for pollinator attraction in food deceptive systems. Therefore, investigating this form of pollination strategy in orchids relies on both observational and behavioral experiments of potential pollinators and their foraging habits. The insects attracted to food-advertising and deceptive orchids can be assessed using the following methods (Figure 8.6):

1. *Direct observation.* Observations of insects visiting orchids during the peak flowering period can be made to identify potential pollinators. This process takes patience as visitation rates are often low, particularly in long-flowering species such as *Paphiopedilum*. During this time, the identity and behavior of insects, the presence and abundance of orchid pollinia, weather conditions, and time of day should be recorded. Direct observations should be undertaken at peak periods of activity of the expected pollinators with activity highest in clear, sunny, and warm conditions. Care must be taken if there are risks that pollinator behavior is influenced by observer presence—think bird hides.
2. *Pollinator observations by video recording.* This method provides a good alternative to long days of direct observations and operates continually under any weather conditions. The process is constrained by the cost of motion activated cameras and the fact that one (or only a few) flowers can be recorded at one time (Micheneau et al. 2009), with many cameras limited to organisms of sufficient size to trigger the camera mechanism.
3. *Pan traps for assessing insect diversity and abundance.* These are often yellow-colored plastic dishes filled with water containing a small amount of detergent. Pan traps are placed within orchid patches during the flower period from morning to afternoon. Captured insects can be inspected for pollen and identified as potential pollinators.
4. *Malaise traps for assessing insect diversity and abundance.* These are tent-like gauze structures that act like stationary nets that can be set up in the vicinity of orchid

Figure 8.6 Types of trapping devices to catch insect pollinators: (a) direct observation requires a great deal of patience but under the right conditions can be a very effective method for capturing orchid pollinators. Here, an Adelaide University researcher watches a patch of endangered *Caladenia rigida* in the Mount Lofty Ranges, South Australia; (b) SLAM (Sea, Land, and Air Malaise) traps are a modified version of the standard malaise trap and are easy to install along insect flyways to capture potential orchid pollinators from all four directions; (c) and (d) colored pan traps filled with water and a drop of non-scented detergent are placed within a patch of orchids to actively attract food-seeking insects such as native bees (photos: Renate Faast).

populations to intersect with the natural flyway. As for pan traps, captured insects can be inspected for pollen and identified as potential pollinators.

5. *Presentation stick method.* Presentation of orchid flowers on a wooden stick approximately 1.5 m long with a crossbar of 40 cm at the top (Johnson 2000). At either end of the cross bar, orchid flowers are fixed; one supplemented with artificial nectar, one without. Artificial nectar (sucrose solution) is equivalent to the main plant source of nectar insects would utilize at that site. Flowers are presented during the warmest hours of the day in sunny conditions, usually over a one-week period. Insect behavior is thoroughly monitored, including the initial choice of inflorescence, the time spent on each flower, the total number of flowers visited on an inflorescence, and distances flown by flies after leaving the experimental inflorescences (Jersakova and Johnson 2006).
6. *Insect sweep netting.* Sweep netting using a standard entomological net will capture a variety of insects, hopefully some bearing pollinia, when foraging on nearby non-orchid plants (Dafni and Calder 1987; Indsto et al. 2005) is a reliable and simple approach that can be effective in locating potential pollinators. Insects can be enticed into activity by spraying nearby foliage with a diluted (5–10%) honey solution.

The methods described here, while suitable for determining potential pollinators of orchids, need to be supplemented by direct observation of visitation where both removal and deposition of pollen is observed. Given the rarity of these observations, researchers have employed insect-trapping techniques in conjunction with DNA analysis of pollinia carried by putative pollinating agents (see Widmer et al. (2000) and Indsto et al. (2005) and Chapter 9).

Case Study 8.1

Evaluating the role of flower color signal mimicry in pollinator attraction and pollination success in the slipper orchid Paphiopedilum micranthum

Xiaokai Ma and Yi-bo Luo
State Key Laboratory of Systematic and Evolutionary Botany,
Chinese Academy of Sciences, Beijing

As a flagship group for orchid conservation, slipper orchids (subfamily Cypripedioideae) have been extensively studied in terms of their reproductive strategies since Darwin (1862). All slipper orchids utilize deceptive systems (nonrewarding) and generally employ specific floral signals (e.g., color and scent) to mimic the models (rewarding flowers or aphids that are prey for some *Cypripedium* pollinators) for pollinator attraction, thereby ensuring reproductive success by using the pollination guilds of other plants.

The silver slipper orchid, *Paphiopedilum micranthum*, which relies on bumblebees for pollen transfer, can mimic color signals of sympatric nectar- or pollen-providing flowers (Figure 8.7). To investigate the functional significance of color mimicry in facilitating pollinator attraction and promoting pollination success, we used insect color vision analyses and behavior tests in natural stands of the orchid. Floral color patterns as perceived by pollinators of *P. micranthum*, its congeners, and sympatric rewarding flowers were measured by plotting the reflectance of the outer and central parts of the flower as separate loci in a bee vision model (color hexagon, Figure 8.8) suggested by Chittka (1992). The colors are divided into six partitions (B,

Figure 8.7 Pollinator bumblebee *Pyrobombus flavescens* visiting *Paphiopedilum micranthum* and sympatric rewarding flowers (from Ma et al. 2016).

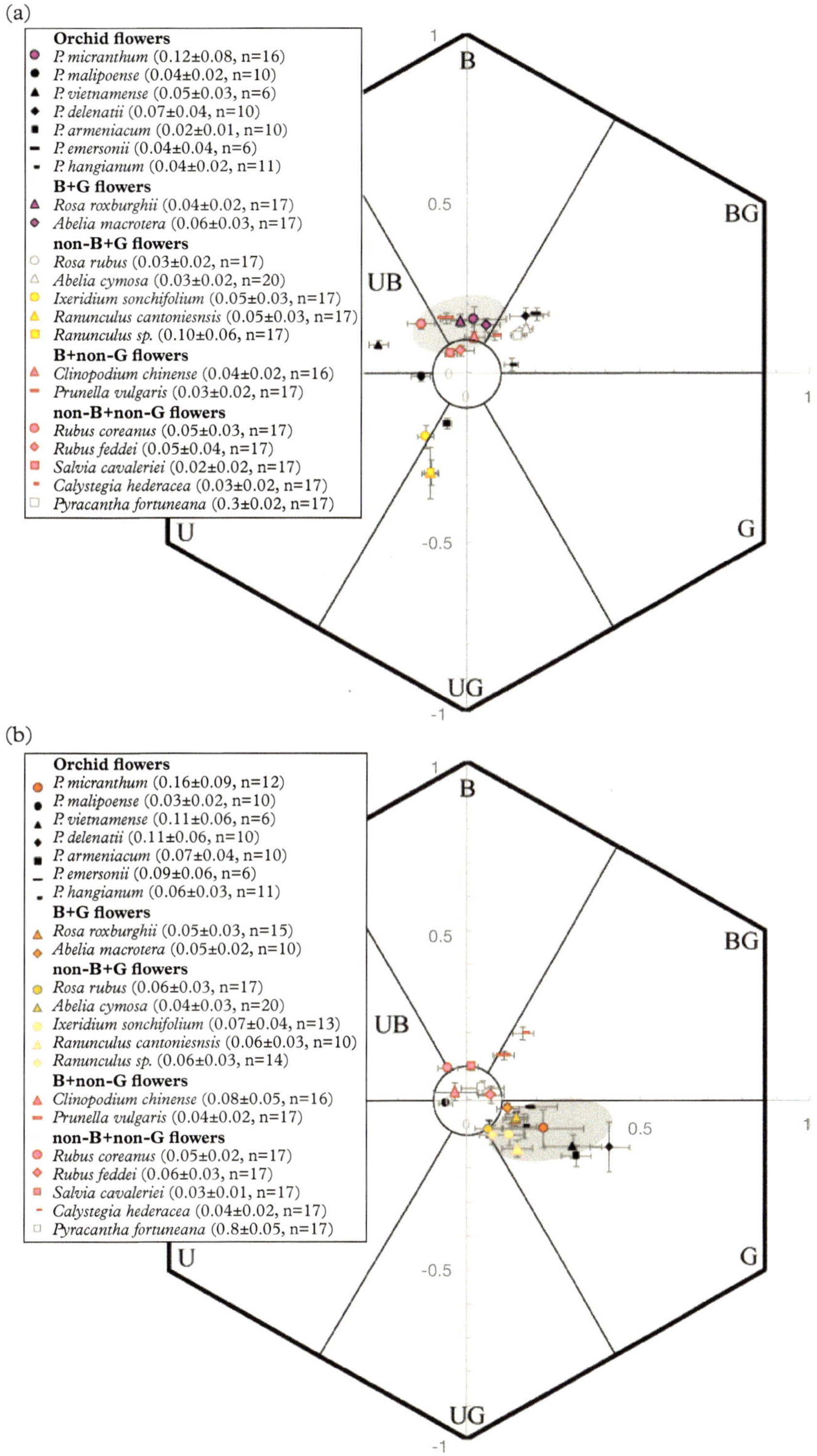

Figure 8.8 Color loci of *Paphiopedilum micranthum*, its congeners and sympatric food flowers in the bee color hexagon proposed by Chittka (1992) as a key to discrimination by pollinators. (a) Loci of outer color of flowers and (b) Loci of central color of flowers (Ma et al. 2015).

BG, G, UG, U, and UB) and one uncolored central circle (C), according to the colors perceived by bees. After that, the effect of food flower color patterns was tested using a bee *interview* approach (Johnson et al. 2003). In addition, the flower color pattern was manipulated using paper discs, and its effects on orchid pollen transfer were recorded (Figures 8.9 and 8.10).

Our results showed the probability of both a long-range approach and a nearby choice by bumblebee pollinators was enhanced with food flowers that shared a similar color pattern to the slipper orchid. Probability of the long-range approach was negatively correlated with color distance between orchid and food flowers. Flowers of *P. micranthum* that were manipulated to reduce complexity of color pattern had reduced pollen removal and lower pollination success. Our study highlights the functional significance of complex floral signals for orchid success and the importance for considering species-interactions in devising orchid conservation strategies that need to also conserve and enhance the specific pollinators (Ma et al. 2015).

Figure 8.9 Manipulated orchids used in behavior experiments. (a) Intact *Paphiopedilum micranthum* flower; (b) *P. micranthum* flower with staminode emasculated; (c) *P. micranthum* flower with staminode emasculated and yellow disc added; (d) with green disc added; (e) with ultraviolet-reflective yellow disc added; (f) pollinator *Pyrobombus flavescens* approaching the orchid with the central yellow disc; and (g) carrying pollinia from the orchid (from Ma et al. 2016).

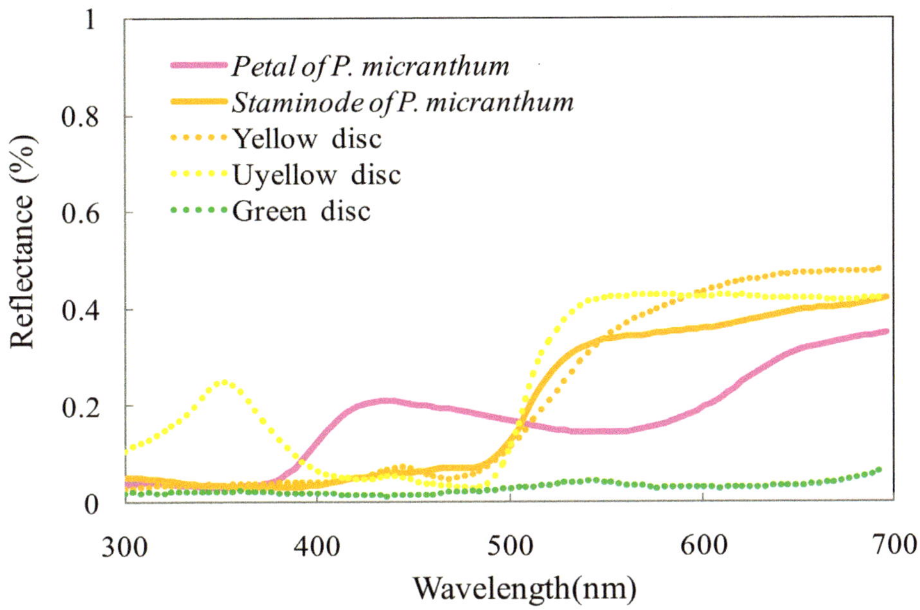

Figure 8.10 Reflectance (%) of outer petal and central staminode of *Paphiopedilum micranthum* flower, and paper discs (Yellow, Uyellow: ultraviolet-reflective yellow, and Green) used in orchid manipulation experiments (from Ma et al. 2016).

8.3.4 Sexual Deception

A less common form of pollinator attraction involves sexual deception. Here, flowers mimic the mating signals of female insects, both as floral odors and through visual mimics of females sometimes coupled with tactile (spring-loaded hinges) cues (Cozzolino and Widmer 2005b; Schiestl 2005). While floral odor serves as a long-range attractant, visual and tactile cues appear to be important in the positioning of animals on the flower to achieve pollination. Male insects deceived into sexual behavior with orchid flowers inadvertently collect or deliver the pollinia and do not receive floral rewards such as nectar or pollen (Jersakova et al. 2006; Schiestl 2005). For example, *Ophrys* flowers produce chemical signals that elicit sexual behavior in males, whereby the males attempt copulation with the flower labellum (Figure 8.11). During this pseudocopulatory event, flower pollination is effected (Ayasse et al. 2000).

Sexual deception has evolved independently on multiple occasions in orchids (Cozzolino and Widmer 2005b) with at least 400 species of SD orchids known in major genera, including *Ophrys* in Europe (Borg-Karlson 1990), *Disa* in southern Africa (Steiner et al. 1994), and several South American (Singer 2002; Singer et al. 2004) and central American (Blanco and Barboza 2005) genera. In Australian terrestrial orchids, the strategy of sexual deception is well developed in multiple species across nine genera.

Solitary bees and wasps are the most common pollinators of SD orchids. Such specialized pollination by wasps is rare in almost all other groups of flowering plants, except figs (Johnson 2005). Thynnine wasps are the pollinators of 70 species represented in six of the

Figure 8.11 Sexually deceptive orchids (a) *Ophyrs insectifera* (photo: Julienne Schiebold); (b) *Caladenia discoidea*; (c) *Drakaea livida*; and (d) *Caladenia infundibularis* (photos by Eric Swarts).

nine Australian genera reported to be SD (Gaskett 2011). Ichneumonid and scoliid wasps also pollinate species from Australian SD orchid genera, and sphecid and pompilid wasps pollinate both South African *Disa* species. European *Ophrys* species are predominantly pollinated by solitary bees in the families Andrenidae, Colletidae, Megachilidae, and Apidae. Until recently, orchids were the only plant family known to engage in the sexual deception strategy.

As a result of this pollinator-driven selection, orchids are highly specialized in their floral morphology, scent, color, shape and chemical signaling (Cozzolino and Widmer 2005a; Peakall and Handel 1993; Schiestl and Schluter 2009).

8.3.4.1 Methods for Assessing Sexual Deception and Pollinator Specificity in Orchids

Sexual deception is a highly specialized pollination mechanism where, in the species investigated, each orchid primarily relies on a single species of pollinator. A simple guide is if only male insects visit an orchid flower and the flower is dull colored (though not always the case, e.g., *Ophrys* can be brightly colored) or highly reduced in terms of floral display, then pollination by sexual deception may be an option. There are some cases where a second pollinator has been reported (Phillips et al. 2015) and other cases where orchids share pollinators (Gaskett et al. 2016). Investigation of the sexual deception pollination mechanism has provided scientists with significant insight into how pollinators and their behavior might influence orchid evolution (Text Box 8.3). Pollinator specificity in SD orchids can be demonstrated experimentally via behavioral-choice tests where potted flowering plants or collected flowers are presented in the field (Bower 1996; Paulus and Gack 1990). This method enables assessment of the presence and abundance of pollinators at a site.

8.3.4.2 *Floral Baiting*

Flowers of SD orchids should be exposed in the field, with their stems in small glass vials of water placed on the ground at about the same height as the plant in the wild and in a suitable habitat. Flowering orchids in pots (see Chapters 4 and 6) are an ideal bait flower source as the flowers remain fresh for longer and pots can be easily moved—noting that potted orchids need to be free of pests and disease.

Three flowers from the same population of each orchid species are placed in each vial. Up to eight sets of flowers placed in a line, 35–40 cm apart can then be monitored simultaneously in choice experiments (Figure 8.12). The line of vials should be oriented at right angles to the wind direction to minimize overlap of allomone trails. Flowers should be exposed for a standard of three minutes at each site because most males arrive within the first few minutes of presentation (Peakall 1990b). Relocating flowers to a new site 50 m or more away can obtain a renewed response by sampling a new component of the insect population.

Three comparative field test designs using the methodology just mentioned can be used to answer the conservation questions regarding preferences of particular pollinator species for specific SD orchid taxa (for more details see Bower 1996):

1. *Sympatric choice test.* Two or more sets of flowers thought to possibly represent different taxa from the same geographic region are exposed together. This should be done within and beyond the geographical range of the study orchids to investigate pollinator specificity, abundance, and distribution.
2. *Allopatric choice test.* Samples of orchid taxa from different geographic regions are exposed together in each of their regions and habitats. This is to test for pollinator isolation, which may evolve in allopatry. Different suites of pollinators exert a

Figure 8.12 (a) Setting up bait flowers; (b) To test whether orchid species use sexual deception to lure male insect pollinators, bait flowers such as these *Caladenia rigida* can be artificially presented away from the main orchid population and relocated every 5–10 minutes (photos: Renate Faast).

divergent selection on floral morphologies, such that when the taxa become sympatric, pollinators discriminate between them (a distinct possibility for taxa with habitat isolation).

3. *Sequential test.* This test controls for possible confounding variation in attractiveness to wasps among samples of the same species. It aims to determine whether nil responses to flowers in previous choice tests were due to the greater attractiveness of other conspecific flower samples. It consists of: (a) a three-minute exposure of flower samples with low or nil responses in a previous choice test, followed at the same site by (b) a further three-minute exposure in which previously attractive flowers are added to the array as controls. Responses by wasps in the first three minutes to previously unattractive flowers show they have similar allomones and may be the same species under the biological species concept as the more attractive samples. If the first-exposed flowers remain unattractive and there is a response to the control, the two samples are considered to represent different biological species.

8.3.4.3 *Pollinator Criteria*

Individuals of insect species responding to orchid flowers can exhibit a graded range of behaviors. This is likely to reflect the degree to which the chemical and floral mimicry of the female has succeeded (Bower 1996; Peakall 1990a). Some individuals within a pollinator species may show weak responses to flowers, such as brief visits without landing, whereas others may undergo prolonged mating attempts on the labellum decoy. An insect species is likely to be a pollinator of an orchid species if there is deposition or contact with the stigma, but previous studies have shown that when wasps are observed with pollen or attempting copulation, continued observation has revealed them to be a pollinator (Bower 1996).

Text Box 8.3

Research approaches for investigating sexual deception in orchids

Gaskett (2011) highlighted four research paths for resolving sexual deception in orchids noting that findings from each research approach must be integrated:

1. Natural history observations where the pollinator's sexual behaviors are observed and documented (e.g., Bower 1996; Peakall 1990b).
2. Sensory ecology where field behavioral observations of orchid floral scent, color, shape, and texture are matched with chemical analysis of pheromone signaling (the chemical ecology of orchid pollinators is described in more detail in a later section) and choice tests.
3. Pollinator ecology where pollinator behavior and learning are investigated (e.g., Ayasse et al. 2000).
4. Molecular ecology approach where phylogenetic comparisons are made between orchids and their pollinators, pollinator chemistry, and pollinator behaviors (see Chapter 9).

Case Study 8.2

Pollination by sexual deception

Colin Bower
FloraSearch, NSW, Australia

A remarkable characteristic of SD orchids is extreme pollinator specificity such that each orchid species has evolved to attract a single pollinator species. The attraction is mediated through the orchid's mimicry of the unique sex pheromone odors emitted by the female insect to attract conspecific males for mating (Bohman et al. 2016). So powerful is the attraction, that SD orchid flowers can be used as *bait* to rapidly collect samples of pollinators in the field (Peakall 1990b). By baiting simultaneously with samples of two or more SD orchid species in *choice experiments*, it can be shown that related SD orchids do not attract each other's pollinators (Bower 1996).

Comparative baiting tests in Australia have also revealed the presence of cryptic pollinator-specific orchids, namely, orchid taxa with virtually identical morphology that consistently attract different pollinator species; for example, in the *Chiloglottis valida* complex (Bower 2006; Bower and Brown 2009) (Figure 8.13). *Chiloglottis pluricallata* was segregated from *C. valida* by pollinator specificity and correlated morphological differences (Jones 1991). Subsequently, baiting trials with populations assigned by Jones (1991) to *C. pluricallata* have revealed four pollinator-specific biological species; three on the New South Wales (NSW) Northern Tablelands and one on the Central and Southern Tablelands of NSW extending into Victoria (Bower and Brown 2009). A widespread Northern Tablelands montane forest taxon co-occurs with a

Figure 8.13 (a) *Chiloglottis valida* and (b) *Chiloglottis pluricallata* (Central Tablelands) are morphologically similar species that may grow in mixed colonies on the NSW Central and Southern Tablelands. They can be distinguished by their consistent attraction of different male thynnid wasps for pollination. The *C. valida* pollinator (*Neozeleboria monticola*) is a black wasp with yellow mandibles, while the *C. pluricallata* pollinator (*Neozeleboria nr. impatiens*) has yellow spots on the sides of the abdomen and more extensive facial and thoracic yellowing (photos: Colin Bower).

localized wet forest taxon at Point Lookout in New England National Park and a second localized taxon apparently confined to swamp margins at Barrington Tops. A third widespread taxon on the NSW Central and Southern Tablelands is sympatric with *C. valida*. These biological species have yet to be formally described.

Conversely, comparative baiting trials have also shown that groups of taxonomic species—for example the *Caladenia reticulata* and *Caladenia parva* complexes—may share the same pollinators and belong to the same biological species (Swarts et al. 2014). Testing of hypotheses about cryptic species and pollinator sharing requires the use of robust experimental designs for field choice tests that can distinguish alternative hypotheses (Bower 2006).

Pollinator relationships can help to understand complexes of closely-related taxonomic species, such as the *Caladenia reticulata* complex which contains almost 30 described species, seven of which are listed as critically endangered under Australian law. Pollinator choice testing has shown the complex contains a much smaller number of biological species, as few as six or seven, with one group of at least ten species, including five of the threatened species, all sharing the same common, widespread pollinator species (Swarts et al. 2014). This suggests it may be better to bias the allocation of scarce conservation resources toward biological species, rather than to each of the threatened taxonomic species sharing the same pollinator.

8.4 POLLINATOR INFERENCES IN CONSERVATION PLANNING

First and foremost, for conservation planning, it is important to know the identity and number of organisms functioning as pollinators. Ideally, these questions should be tested at multiple sites to take into account the potential for variation in pollinator species over the geographic extent of the orchid. Resolution of the number of pollinating animals and their identity is often difficult to achieve—as many orchid pollinators, particularly if they are invertebrates, are difficult to identify without specialist advice.

The next stage in conservation planning with pollinators is to know if the abundance, seasonality, and distribution of the insect is limiting the orchid distribution. As orchids are usually dependant on their pollinators for sexual reproduction, orchid distribution must match the distribution of its pollinators. What is the distribution and abundance of the insect, first locally, and then at the larger landscape scale? Does the insect occur in a suitable habitat and sites with the orchid's distribution? If a restored habitat, are the correct food and nest plants available to support the pollinator? The answers to these questions will help determine if the pollinator is limiting existing populations (i.e., through poor fruit set) and if pollinator absence explains the lack of the orchid in an otherwise suitable habitat.

8.5 POLLINATOR CHEMISTRY

Pollination success often depends upon the attractiveness of volatile compounds emitted by the plant to attract specific pollinators. Volatile substances are relatively small molecules with great communicative power for attracting pollinators, recruiting help, and repelling predators. The volatiles, which are composed of several molecules, lure the pollinators to the flowers by advertising food reward(s) or in the case of SD orchids, mimicry—the promise of a mate.

Though there is a large number and wide variety of animal pollinated plants relying on floral volatiles for pollinator attraction, only recently have scientists begun to understand the complex chemistry involved. In SD orchids, a combination of ethological and chemical investigations have shown that floral volatiles are responsible for the attraction of specific pollinators (Ayasse et al. 2003; Bower 1996; Schiestl et al. 2000). In these studies, floral volatiles are considered to be one of the key traits responsible for reproductive isolation of sympatric species and may have played a major role in the evolutionary dynamics of SD orchid lineages (Mant et al. 2005).

The most powerful studies in deciphering the detection of floral odor by pollinators combine behavioral observations and assays with chemical analyses and tests for pollinator perception of natural and synthesized scents. The most common way in which pollinator perception of orchid scent can be tested is through gas chromatography-electroantennography or gas chromatography-electroantennal detection (GC-EAG or GC-EAD; (Schiestl 2005)).

GC-EAD is an analytical procedure that permits the identification of compounds in complex mixtures that stimulate the olfactory sensilla of an insect. This information can be used to discover potentially attractive compounds (such as sex pheromones) that alter the behavior of insects. Using excised insect antennae, the effluent from the GC-EAD is passed across the antenna and electrical impulses are correlated with peaks of chemical compounds appearing on the GC readout. Using this approach, GC output can be used to confirm the presence, identity, and quantity of compounds exposed to the antenna while the antennal (EAD) output indicates presence/absence of olfactory sensitivity to eluting compounds and provides a relative measure of the intensity of olfactory stimulation.

GC-EAD allows scent compounds to be individually tested for their capacity to stimulate a pollinator's antennal chemoreceptors. Whole live antennae of the pollinator species have been tested with synthetic versions of compounds identified in orchids or more commonly, with extracts of orchid labellae and/or females of the pollinator species. For examples specific to orchids where this method has been used, see the following: Schiestl et al. (2000), Schiestl and Ayasse (2002), Schiestl and Marion-Poll (2002), Ayasse et al. (2003), Stokl et al. (2005 and 2007), and Vereecken and Schiestl (2008).

Not every electrophysiologically active compound elicits a behavioral response from the pollinator so behavioral assays must be performed using a synthesized version of the potential chemical displayed on a black pin head (Schiestl and Marion-Poll 2002). The model example of this procedure is the single novel compound attracting pollinators to *Chiloglottis trapeziformis*, and the use of this compound and analogs in further investigations of pollinator specificity in other *Chiloglottis* species (Franke et al. 2009; Poldy et al. 2005; Schiestl and Peakall 2005; Schiestl et al. 2003).

In *Chiloglottis*, field bioassays using synthetic semiochemicals have demonstrated the value of chemistry in pollinator attraction and indicate that two distinct mechanisms control specific pollinator attraction:

- Pollinator attraction by a single specific compound.
- Pollinator attraction by two compounds where only a particular blend triggers specific attraction.

It is predicted that these same two mechanisms for controlling pollinator specificity is likely to be found in other species pollinated by sexual deception.

Text Box 8.4

Future conservation research for orchid pollinator relationships

A key area in orchid pollinator conservation is to develop translocation or captive breeding programs to reinstate or reinforce pollinator capability for the orchid. Pollinators have been identified as a key *missing link* in ecological restoration and plant conservation programs involving reintroduction (Dixon 2009; Menz et al. 2011). At present, there are no cases where pollinators have been artificially manipulated or captive-bred and released in support of aiding orchid pollination. The technology and resourcing of these approaches is unknown, but for specialized pollination syndromes such as the thynnine wasp in Australian orchids, the dearth of information on pollinator life cycle requirements (host range, trophic support, range capability) may require significant research investment.

9

GENETIC TECHNIQUES IN ORCHID CONSERVATION

9.1 BACKGROUND

Conservation genetics provides the theoretical framework and practical tools to assess and conserve diversity, and protect evolutionary capacity in natural and restored populations. The advent of the molecular age has revolutionized the scale, depth, and applicability of genetic knowledge with results available in hours or days rather than months. Conservation or restoration of genetic diversity (and therefore evolutionary potential) is now viewed as a major goal of conservation for rare and threatened species, resulting in the use of molecular methods and phylogenetic frameworks in the design and application of conservation strategies (Falk and Holsinger 1991; Hogbin et al. 2000; Hopper 2000). A rapid expansion in the array of molecular techniques available in the field of orchid conservation has been seen in the last three decades.

Molecular data provides an empirical framework through which the objective quantification of biodiversity can be achieved, helping conservation practitioners define priorities and optimize management decisions (Qamaruz-Zaman et al. 1998). Molecular data, using a multiplicity of techniques, provides vital information from genes to the population (population genetics) and species (phylogenetics) level. This chapter summarizes molecular techniques used in orchid conservation from both a population genetic and phylogenetic perspective and details how genetic data can inform population management and priority assignment (Figure 9.1).

9.2 POPULATION GENETICS

Population genetic methods provide the basis for assessing the evolutionary scenarios/history underlying the genetic makeup of a population that, in turn, provides the ability to understand impacts on the genetic structure of plant loss. In this way, conservation practitioners can prioritize populations/individuals for conservation or identify the most appropriate plants for long-term storage or use in reintroduction programs. As research continues on the genetic diversity/structure of orchid populations, the closer we can get to deriving and predicting patterns to facilitate the decision-making process.

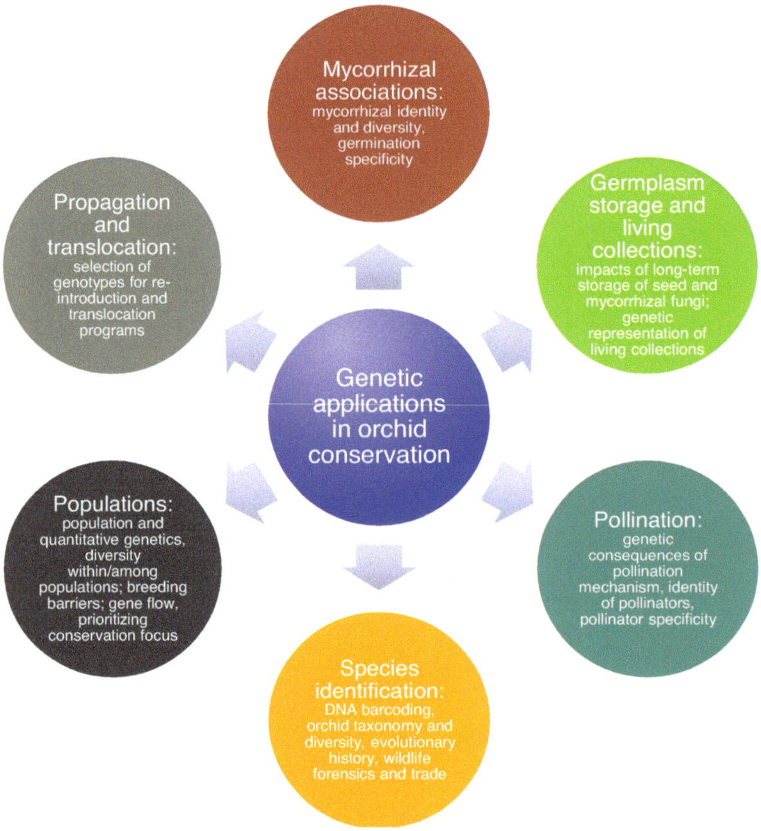

Figure 9.1 Genetic applications in orchid conservation

Population level studies aim to survey the genotypic range within and among populations. Molecular data, in this context, can be used to determine:

- If populations are distinct and may warrant being treated as separate conservation units;
- The level of variation is within and among populations to identify important plants and populations; and
- The genetic substructure of populations to ensure translocation programs achieve some level of genetic matching with the reference population.

Following this, a range of analytical approaches have been developed to elucidate the effects of genetic bottlenecks, gene flow, inbreeding, and genetic drift. In orchid research, these methods are being used to study the impacts of habitat fragmentation on gene flow and population structure (Wallace 2002), genetic consequences of pollination syndromes (Peakall and Beattie 1996), evolutionary potential of extant populations (Ackerman and Ward 1999), cost of inbreeding (Wallace 2003), and fine-scale spatial structure in orchid populations (Chung et al. 2005).

Qamaruz-Zamen et al. (1998) and Fay and Krauss (2003) detail the variety of methods used in population genetic studies in orchids. Molecular markers that are codominant, single-locus markers are those that give the most useful and reliable estimates of genetic information. These include allozymes, restriction fragment length polymorphisms (RFLPs), amplified fragment length polymorphisms (AFLPs), microsatellites, and DNA sequencing. Indeed, much of our knowledge of orchid population genetics has come from allozyme studies (Brzosko et al. 2002; Chung and Chung 1999; Sun and Wong 2001), and with the increasing number of orchid specific microsatellite primers available, studies employing these highly informative markers are on the increase (Cozzolino et al. 2003; Gustafsson and Lonn 2003; Lombardo et al. 2008; Pellegrino et al. 2007; Swarts et al. 2009).

Traits in orchids—such as deceptive pollination strategies and dust-like, predominantly wind dispersed seed—promote outbreeding and gene flow between populations. Such traits are associated with high levels of diversity within populations, and lower genetic differentiation between populations (Hamrick and Godt 1996). While this pattern of population genetic structure is often reported for orchids (Gustafsson 2000; Machon et al. 2003; Soliva and Widmer 2003), other studies have documented a wide range in the levels of genetic diversity and differentiation between orchid populations (e.g., Wallace 2002; Wong and Sun 1999). Significant heterogeneity in the genetic structure of orchid populations was documented by Phillips et al. (2012) based on a meta-analysis of 52 allozyme studies and is one of the lowest reported in plants and the lowest for any herbaceous family. Data based on microsatellite analysis of the Orchidaceae showed a similar trend (Phillips et al. 2012), but there were only limited studies highlighting the need for population genetic analysis using these markers within the orchids.

Text Box 9.1

Polymerase chain reaction

The polymerase chain reaction (PCR) is a foundation molecular technique that amplifies specific segments of DNA across several orders of magnitude, potentially generating thousands to millions of copies of a particular DNA sequence from a single DNA template. The method relies on thermal cycling (Figure 9.2a), consisting of cycles of repeated heating and cooling of the reaction for DNA melting and enzymatic replication of the DNA (Figure 9.2b). Primers (short DNA fragments) containing sequences complementary to the target region along with a DNA polymerase are key components to enable selective and repeated amplification. Included in the method are dNTPs as building blocks for new DNA, a buffer for creating optimal reaction conditions for Taq, and salt for DNA stabilization. As a PCR progresses, the DNA generated is itself used as a template for replication, setting in motion a chain reaction in which the DNA template is exponentially amplified. PCR can be extensively modified to perform a wide array of genetic manipulations, allowing visualization on an electrophoresis gel.

Four steps are involved:

1. Denature: The reaction mix is heated to 94–98°C to separate the double stranded DNA template into single stranded molecules.
2. Annealing: Lower temperature (40–70°C) to allow short DNA molecules called *primers* (8 to 30 base pairs in length) to *anneal* to their complementary strands.

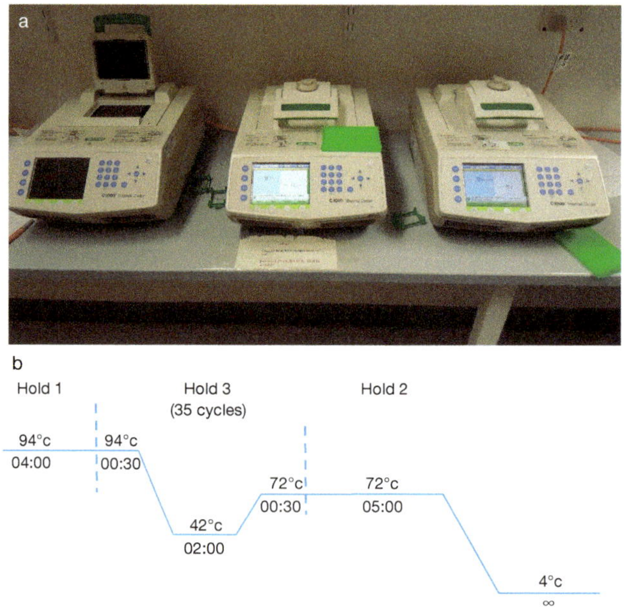

Figure 9.2 (a) Thermocycler to complete the polymerase chain reaction (PCR) and (b) Thermocycling profile for PCR. The use of heat stable DNA polymerases which survive exposure to high temperatures required by the PCR, and the development of thermocyclers capable of cycling temperatures quickly and accurately have facilitated the automation of this process and revolutionized the analysis of genetic variation at the population level.

3. Extension: Increase temperature to 72°C to allow the copying of single-stranded DNA commencing at the primer-template complexes. DNA polymerase is the catalyst for this extension.
4. Repeat steps 1–3 for 20 to 50 cycles to amplify the regions of DNA bounded by the primers exponentially.

Text Box 9.2

Codominant and dominant markers for conservation genetics

There is a wide array of markers for conservation genetics; however, these can be split into two main categories:

Codominant marker—a marker that visualizes both alleles in a diploid organism to detect homozygous and heterozygous genotypes. Examples of genetic markers include allozymes, cleaved amplified polymorphic sequence (CAPS), RFLPs, and microsatellites (genomic single sequence repeats (G-SSR)), and more recently with cheaper sequencing technology, expressed sequence tag-SSR based on transcript sequences, which produce a distinct pattern of bands on a gel for each genotype at a given locus.

Dominant marker—a marker that visualizes only one allele in a diploid organism where only two genotypes are detected. Heterozygous individuals are indistinguishable from one of the homozygous types. Examples include Random Amplified Polymorphic DNA (RAPD) and inter simple sequence repeat (ISSR); directly amplified minisatellite DNA (DAMD); and Start Codon Targeted (SCoT) and AFLP markers, which are based on either band presence or band absence, with the gene product of only one allele (in either homozygous or heterozygous form) detected.

Of these, a distinction can be made between *PCR* and *non-PCR-based techniques.* Non-PCR techniques include allozymes and RFLP. PCR techniques include RAPD, AFLP, and microsatellites. Refer to Section 9.7 for scoring of these data sets.

9.2.1 Principal Population Genetic Markers for Conservation

Allozymes: Alternate forms of enzymes encoded by different alleles at the same locus. Allozymes are prepared by homogenizing tissue to produce a solution of proteins that is electrophoresed through a gel (starch or cellulose acetate). Specific enzyme products are then visualized by the use of a stain facilitating the enzymatic reaction. For example, alcohol dehydrogenase is scored by the use of alcohol, which is converted to aldehyde that binds with a dye for the visualization of bands. Alleles of different sizes have different mobilities.

RFLPs: Polymorphisms at specific sites in the DNA sequence revealed by cutting DNA with restriction enzymes, electrophoresis, and blotting these products to a membrane which is then probed with radioactive DNA allowing the visualization of bands of different size.

RAPDs: Polymorphisms obtained by the amplification of DNA fragments by PCR using *arbitrary* primers. The primers are typically 10 bases long and serve as both forward and reverse amplification. The DNA fragments are electrophoresed in an agarose gel and visualized with the use of ethidium bromide and an ultraviolet (UV) light source.

Microsatellites (or SSRs): Tandem repeats of very short nucleotide motifs (1–6 bases long) [e.g., (CA)17 or (AAG)12]. Specific primers amplify these sites by PCR, with visualization by electrophoresis on agarose or acrylamide gels using an automated sequencer and fluorescent labeling. These markers often present high levels of inter- and intra-specific polymorphism, particularly when tandem repeats number ten or greater. Microsatellite loci are widely distributed throughout the genome and can be isolated from semi-degraded DNA of aging or dried samples, as all that is needed is a suitable substrate for amplification through PCR. Microsatellites developed for particular species can often be applied to closely-related species, but the percentage of loci that successfully amplify may decrease with genetic distance.

AFLPs: Markers using a powerful PCR-based multi-locus DNA fingerprinting technique with high reproducibility and wide application. Steps involved are:

1. Restriction of the DNA with restriction enzymes such as EcoRI and MseI;
2. Ligation of adapters to the ends of these fragments;
3. A pre-selective PCR, using an EcoRI primer and an MseI primer, each with a 1 base extension beyond the adapter site;
4. A selective PCR amplification, using EcoRI and MseI primers with a 2, 3, or 4 base extension beyond the adapter site—the EcoRI primer is fluorescently labeled, enabling electrophoresis and visualization on an automated DNA sequencer; and

5. Scoring of fingerprints using dedicated software such as GeneScan or Genotyper. Individual fragments are accurately sized by the inclusion of an internal size standard of a different wavelength (color).

9.3 MOLECULAR PHYLOGENETICS/SYSTEMATICS

Phylogenetic data provides the tools for studying relationships between species at multiple scales—below and above species rank. This information relies on molecular DNA sequencing data and matrices generated by interrogation of a wide variety of morphological trait data.

A phylogenetic tree contains nodes that represent a taxonomic unit (a species, a population, or an individual) and branches that define the relationships between the nodes in terms of plant ancestry. The branches are often varied in length or scaled, representing the number of changes that have occurred along that branch. Molecular phylogeny attempts to use these trees to determine the rates and patterns of change occurring in DNA and proteins and to reconstruct the evolutionary history of genes and organisms. The rapid accumulation of sequence data that is now possible from genes to entire genomes is a major advance with significant advantages over studies that rely only on morphological data. Sequences can be generated from nucleotide, mitochondrial, and chloroplast DNA, as well as proteins. The DNA or protein sequences generated are then analyzed for variation and used to depict the relationship of genes and the organism.

As well as the intra-familial role phylogenetics/systematics plays in conservation, an ever-increasing role of phylogenetics is its use in developing and testing species concepts. The phylogenetic species concept and its three derivatives—(1) diagnosable, (2) monophyly, and (3) diagnosable and monophyly (see Pridgeon 2003 for a description of modern species concepts)—was developed to produce a lineage-based concept without the assumptions of certain processes, such as reproductive isolation. The question asked of these species concepts is whether the smallest taxonomic unit studied is diagnosable, monophyletic, or both.

Due to the considerable interest in the Orchidaceae from a taxonomic and evolutionary perspective, the family is quickly becoming the most intensively studied plant family in terms of phylogenetic relationships. In addition, due to their diversity of form, epiphytism, intimate relationships with mycorrhizal fungi, and complex pollination strategies, they represent a premier system in which to study such phenomena.

The increasing number and availability of informative genes used for phylogenetic reconstruction, particularly at the higher taxonomic levels, have significantly advanced the knowledge of orchid classification and stimulated the proposition of new hypotheses by leaders in the field (Chase et al. 2003). Perhaps most important for orchids, these data highlight phylogenetically informative morphological characters partitioned within and among clades. This process, therefore, highlights characters both in need of further study from a developmental/evolutionary perspective and those that are significant from a conservation perspective.

As conservation practitioners, we realize the rate of change (climate, habitat degradation, hydrology) is now one of the key drivers in the decision-making process—delay and a habitat or species may be lost. So how can phylogenetics assist and inform good decision making? Phylogenetics provides us with tools to understand patterns within clades, determine evolutionary histories, highlight divergent individuals, and thus, can be used to fast track decisions needed in conservation planning and management (see Figure 9.3).

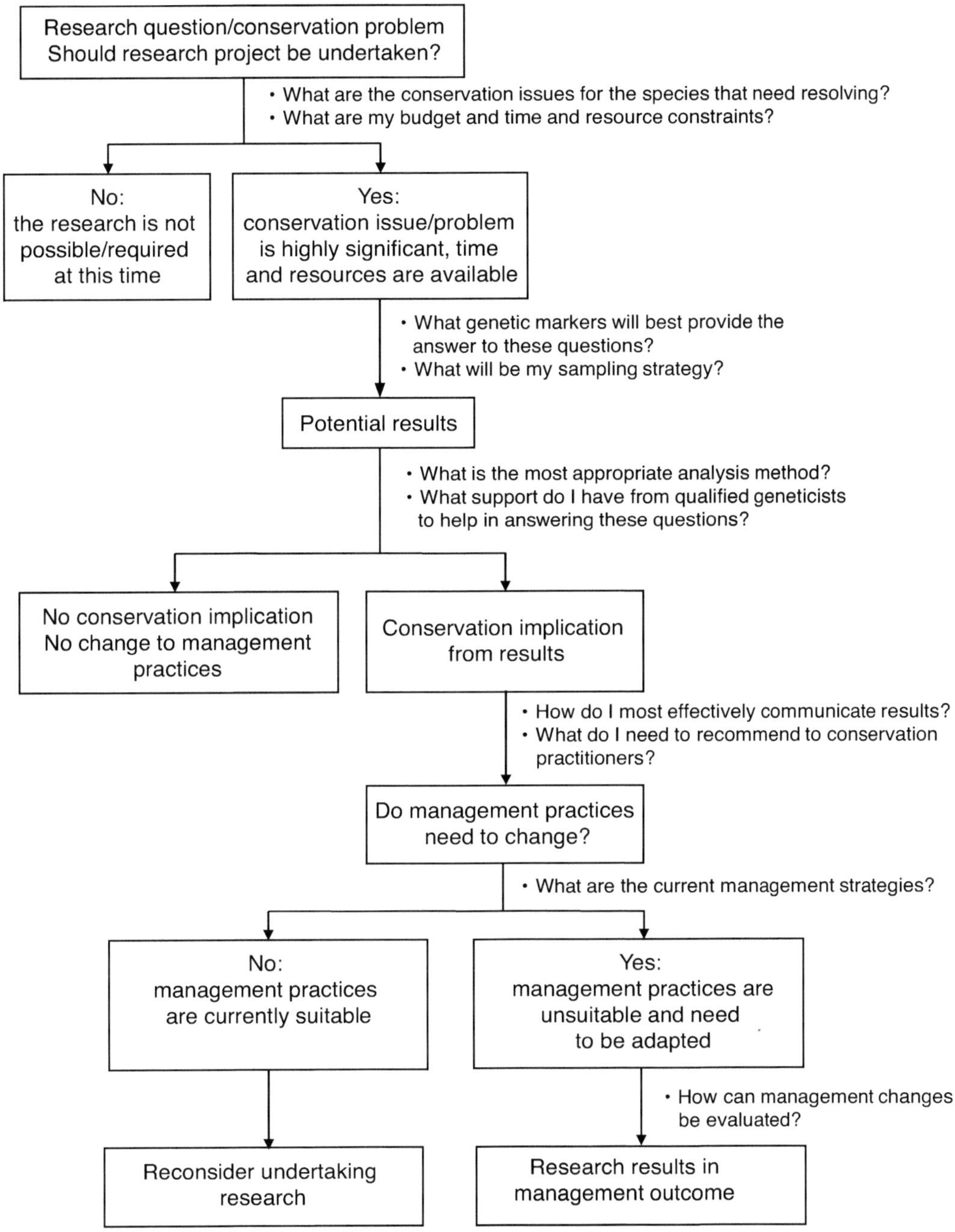

Figure 9.3 A decision matrix to guide conservation practitioners on the practical merits of undertaking a genetics research program for deriving conservation-relevant information

Case Study 9.1

The taxonomically complex genus Ophrys

(Devey et al 2008)

For conservation practitioners, a stable nomenclature is highly desirable to facilitate conservation planning and programs. However, species delimitation in orchids is highly controversial as species groups can be difficult to separate morphologically, and both professionals and citizen scientists debate as to what constitutes a species. The high-profile European orchid genus *Ophrys* is the subject of significant debate and is a genus characterized by highly variable, yet possibly convergent morphology, with a lack of sequence variation that has hindered production of a robust phylogenetic framework.

In *Ophrys* taxa, it is well understood that the pollination syndrome of sexual deception has contributed strongly to the complex morphology of the flower (see Figure 9.4). The insect-like labellum mimics the female wasp through a combination of chemical and visual stimuli in order to attract males of the same species. Because of these morphological complexities, centuries of morphological studies have produced monographs with extremely large variations in the number of species comprising the genus.

In their study of *Ophrys*, Devey et al. (2008) aimed to produce a phylogenetic framework as the basis for more rigorous species delimitation by using a suite of DNA-based phylogenetic and population genetic techniques. Nuclear and plastid DNA sequencing and AFLPs were performed on 85 accessions of *Ophrys*, spanning the full range of species aggregates currently recognized. A combination of parsimony and Bayesian tree-building techniques and principal coordinate analysis were used to analyze the complex data set.

Figure 9.4 (a) *Ophrys apifera* (photo: Julienne Schiebold); (b) *Ophrys bertolonii* (photo: Jacopo Calevo); (c) *Ophrys scolopax* (photo: Jacopo Calevo); (d) *Ophrys omegaifera* (photo: Craig Williams); (e) *Ophrys holosericea* (photo: Jacopo Calevo); (f) *Ophrys incubacea* (photo: Jacopo Calevo); (g) *Ophrys tenthredinifera* (photo: Pankaj Kumar); and (h) *Ophrys sicula* (photo: Pankaj Kumar).

Through this analysis, new insights into *Ophrys* systematics were uncovered, most notably, the recognition of the novel *O. umbilicata* group, which was shown to be reproductively isolated from the other groups due to strong geographic partitioning. A total of ten genetically distinct groups were included within the genus; six of these proving to be robust under stringent phylogenetic analysis. These groups were further assigned to three recognizable sections, which have previously been treated as subgenera by some authors. It was suggested that putative *Ophrys* species may have arisen through hybridization rather than divergent speciation. This is most likely due to the *pseudocopulatory* pollination syndrome of *Ophrys* being less specific than originally thought, and consequently, the genus has been potentially over-divided at the species level.

Case Study 9.2

Salep—species identification of orchid tubers using molecular techniques

Hugo J. de Boer
Department of Organismal Biology, Evolutionary Biology Centre, Uppsala University, Sweden; Naturalis Biodiversity Center, Leiden, The Netherlands; The Natural History Museum, University of Oslo, Oslo, Norway.

DNA sequencing enables identification of species without the need to rely on the presence of morphological characters (Hebert et al. 2003). In recent years, such DNA-based identification—DNA barcoding—has been used to identify and discern species at any developmental or processing stage from which DNA can be extracted, such as DNA in dung (Hibert et al. 2013), pollen (Richardson et al. 2015), and permafrost preserved subfossils (Willerslev et al. 2014). In plants it has been used for purposes including biodiversity inventories (Thompson and Newmaster 2014), wildlife forensics (Deguilloux et al. 2002), and authentication of herbal products (de Boer et al. 2014; Kool et al. 2012; Newmaster et al. 2013).

DNA barcoding is also relevant for orchids, and can facilitate identification of orchids in trade—especially orchids or parts thereof that are traded in a sterile state or lacking morphological characters for identification. An example of such trade is that of *salep* (Figure 9.5a). In Asia Minor, Iran, and the northeastern Mediterranean region, tubers of terrestrial orchids are collected extensively from the wild (Figure 9.5b, c), where they are traded illegally for use in the preparation of ice cream and a drink, known as *salep* (Ghorbani et al. 2014a; Ghorbani et al. 2014b; Kreziou et al. 2015). It is estimated that 20 million tubers, each representing an individual orchid, are destructively harvested in Turkey each year (Kasparek and Grimm 1999)—and 10 million in Iran (Ghorbani et al. 2014a) (see Figure 9.5). Morphology-based approaches for identification of *salep* tubers are insufficient and cannot even accurately distinguish dried tubers from different genera.

Figure 9.5 (a) Dried orchid tubers from Iran with different morphology showing oval and palmate shapes (photo: Aline Nieman); (b) orchid tubers harvested by local people before washing and cleaning; (c) orchid tubers collected in Golestan province, Iran, washed, ready for drying; (d) *Dactylorhiza umbrosa*, in wet grassland in Ardabil province, Iran; and (e) collection of *Ophrys* tubers in Golestan province, Iran (photos: Abdolbaset Ghorbani).

Ghorbani et al. (in press) used DNA barcoding to identify 150 individual tubers from 31 batches purchased in 12 cities in Iran to assess species diversity in commerce. DNA sequences from these unknown tubers are queried against a reference database of sequences from vouchered and identified orchid species to match and identify individual orchid tubers (Figure 9.5d, e). In practice this meant extracting total DNA of the 150 tubers, and amplifying and sequencing the molecular genetic markers, nrITS, *trn*L-F spacer, and *mat*K for each of them. The reference database consisted of 490 nrITS, *trn*L-F spacer, and *mat*K sequences of 133 taxa, including 252 sequences sourced from the public repository NCBI GenBank and 238 new sequences from collections made in Iran for that study. The markers enabled unambiguous species identification with tree-based methods for nrITS in 67% of the tested tubers, 58% for *trn*L-F, and 59% for *mat*K. Among the sampled tubers, species in the following genera were identified *Orchis* (34%), *Anacamptis* (27%), *Dactylorhiza* (19%), *Himantoglossum* (7%), and *Steveniella* (1%). Their study shows that DNA barcoding is a powerful tool for identification of orchids in trade, and enables evidence-based targeted conservation activities for harvested species in their regions of origin.

9.4 QUANTITATIVE GENETICS

Knowledge of the way in which a plant responds to environmental change is an essential component in a conservation/restoration program. One way to gauge the ability of a population to respond effectively to threats is by understanding underlying quantitative genetic variation, that is, the diversity upon which selection acts (Kramer and Havens 2009). Although for decades researchers have sought to understand how and when quantitative variation and selection interact to drive local adaptation, continued progress using these methods has been hampered by the belief that these methods take significantly more time and resources than molecular studies.

Today, conservation and restoration priorities can be based upon findings from molecular genetic studies using noncoding markers (markers which are selectively neutral or regions of unknown function) as described previously. Although these studies provide useful information for conservation, they have limited capacity to predict conservation success (Storfer 1996), as neutral molecular variation rarely estimates variation in ecologically important traits (continuous traits) that confer survival in a changing environment (quantitative genetic variation). For example, populations with little to no measurable difference in neutral alleles may contain important phenotypic plasticity or genetic diversity in traits such as enhanced dispersal ability. In addition, variation at the protein (allozymes) or DNA level may not necessarily be correlated with the ability of a population to survive in a stochastic environment.

Traditionally, measurement of quantitative genetic traits in plants are taken on phenotypic (visible) morphological or behavioral traits that are continuously distributed in a population. Morphological traits are straightforward to measure and can be taken numerous times within a growing season and over multiple seasons. For conservation purposes, it is important to choose traits that are correlated with survival. Such traits may include plant height, number of flowers, leaf width, and fruit set. Ecophysiological responses (such as response to light) are more complex to assess, particularly because response patterns vary within an individual and are dependent on habitat factors such as light availability, etc. While collecting data for quantitative studies, methods are often non-destructive and do not require material for DNA extraction. However, logistical drawbacks can make these studies unsuitable, as large sample sizes are required and often in the case of rare and endangered taxa, this can be a constraint.

Until recently, it was impractical to identify the genes that are responsible for variation in continuous traits, or to directly observe the effects of their different alleles. Now it is possible to identify the genetic basis of trait differences within and between species, using quantitative trait loci (QTL). These are genetic loci identified through the statistical analysis of complex traits (such as plant height or biomass). These traits are typically affected by more than one gene and the environment. They are regions of a chromosome or, ideally, individual sequence variants that are responsible for trait variation. QTLs can be identified at a molecular level (for example, with AFLP) to help map regions of the genome that contain genes involved in specifying a quantitative trait. There is extensive literature dedicated to the mapping of QTL loci (e.g., Mackay et al. 2009), statistics (Barton and Keightley 2002; Hill 2010) and associated problems (e.g., Weller 2009).

In summary, although quantitative genetic techniques, particularly QTLs, are relatively new to the field of orchid conservation, in time they will provide a valuable tool for understanding orchid evolution, and for determining key traits underpinning orchid survival. Studies of adaptation that relate geographic variation (in amount of genetic variation, for example) to environmental and ecological factors can provide insight into specific environmental factors that contribute to the decline of the species in question. As more and more orchid genomes are resolved, quantitative genetic methods will be able to provide important new information for conservation planning.

9.5 MYCORRHIZAL SPECIFICITY IN ORCHID POPULATIONS

One of the most important issues in the conservation and restoration of orchid populations is the association between orchids and their mycorrhizal fungi. The biology of this association has already been presented in earlier chapters. Traditionally, pure cultures based on isolations from orchid tissues, generally involving single pelotons, were the main source of information about orchid symbionts. Fungi associated with orchids rarely produce sexual (anamorphic) stages in pure culture and ambiguous mycelial characteristics frequently prevent identification at lower taxonomic levels. Hence, the identities of many orchid mycorrhizal (OM) fungi and the levels of specificity have long been poorly understood (Curtis 1939; Masuhara and Katsuya 1994; Rasmussen 2002).

Molecular approaches have greatly improved our ability to identify fungi through DNA sequencing and phylogenetic analysis, particularly at higher taxonomic levels. OM fungi from terrestrial orchids are generally classified as belonging to rhizoctonia-like fungi, a polyphyletic group of fungi from three basidiomycete families (Sebacinaceae, Ceratobasidaceae, and Tulasnellaceae) (Roberts 1999).

Here we define mycorrhizal specificity as the phylogenetic breadth of fungi associating with an orchid plant or species (Taylor et al. 2002). It is important to note that fungal presence in orchid tissues may not equate with the fungus operating as a mycorrhiza. A functional mycorrhizal association can only be determined when a growth or development benefit in the orchid is demonstrated or proven.

From germination through to adulthood, the mycorrhizal relationship for orchids is subject to compatibility barriers (Brundrett 2004; Roberts 1999). In contrast, most other non-orchids are mycorrhizal generalists, exhibiting phylogenetically broad associations. Although overall mycorrhizal specificity in the orchid family is narrow, there is considerable variation in specificity among orchid species. In propagation and reintroduction programs, ensuring the orchid is germinated with compatible and efficacious mycorrhizal fungi has important ramifications. An effective mycorrhizal association influences how the orchid accesses soil, water, and nutrient resources, resists stressors such as disease, copes with inter-plant competition, and facilitates future recruitment opportunities for establishment at a site.

Since peloton formation is an important criterion for an orchid symbiont, DNA analysis based on single pelotons extracted from fresh or dried orchid tissue, with the isolate then grown on nutrient agar media, is the most effective source of mycorrhizal material for DNA extraction. As multiple fungal taxa may be found to associate with orchids, not all are likely to be equally efficacious, thus pure cultures grown on nutrient agar provide the capacity to test the mycorrhizal efficacy of the fungus as germination promoters. Direct DNA extraction from orchid root tissue is particularly useful when a fungus is unable to be cultured, however, the functionality of the mycorrhiza present in the root is at present difficult to determine.

Methods for mycorrhizal diversity and specificity studies are mostly based on PCR with fungal specific primers using cultured fungi or macerated orchid tissue. PCR will amplify selected parts of the fungal genome for further characterization of the DNA, either by sequencing or fragmentation patterns. The level of distinction between strains depends on the variability of the genome component under study. Provided corresponding DNA from suitable identified reference fungi is available (e.g., GenBank), a phylogenetic analysis will place the unknown strain in a taxonomic group. Primers generally recommended for fungal barcoding are the universal eukaryotic internal transcribed spacer (ITS) primers ITS 1 (5′-TCCGTAGGTGAACCTGCGG-3′) and ITS 4 (5′-TCCTCCGCTTATTGATATGC-3′), which span the ITS 1, 5.8S, and ITS-2 regions in the rRNA gene operon (White et al. 1990). While these are ideal for isolated mycorrhizal cultures, they are too general for direct sequencing from orchid roots for mycorrhizal community profiling (Waud et al. 2014). More specific primer pairs have been developed for orchid mycorrhizal research (Taylor and McCormick 2008) targeting key Basidiomycota families of fungi typically known to associate with orchids including Sebacinaceae (Ruibal et al. 2014), Tulasnellaceae (Ruibal et al. 2013; Taylor and McCormick 2008) and Ceratobasidiaceae (Gardes and Bruns 1993).

More recently, mycorrhizal fungi sequences are being grouped into Operational Taxonomic Units (OTUs), defined by 97% sequence identity (Jacquemyn et al. 2011; Phillips et al. 2016; Swarts et al. 2010; Těšitelová et al. 2013). This sequence similarity cutoff provides an effective way of estimating the species richness of mycorrhiza associating with orchids in the absence of morphological characters. This method may, however, over- or underestimate fungal diversity, and research is still required to determine the functional significance of each OTU.

Text Box 9.3

Additional PCR-based methods for orchid mycorrhizal research

Quantitative or real-time PCR (qPCR)

This technique monitors the amplification of target DNA during the PCR process in real time—not at its end, as with conventional PCR. As such, this method is used to determine the exact concentration (relative or absolute) of the amplified DNA in the sample. The amplified DNA is fluorescently labeled and monitored constantly during the PCR cycling. The fluorescent signal released during the process is directly proportional to the amount of amplified DNA. This technique has important application for quantifying microorganisms that cause disease, such as bacteria and viruses. In orchid mycorrhizal research, this technique has been used to demonstrate that orchid seed germination *in situ* can be manipulated by altering the abundance of attendant mycorrhizal fungi (McCormick et al. 2012).

Next-generation sequencing (NGS)

This technique, applied through a variety of technologies, enables the sequencing of millions of small DNA fragments in parallel. High throughput technology and fluorescent labeling of DNA allow for fast and cost effective analysis of whole genomes—deep sequencing of target regions to identify novel RNA variants for genetic expression analysis. Two commonly utilized NGS approaches include 454 amplicon pyrosequencing and Illumina (Solexa) sequencing. Waud et al. (2014) used 454 pyrosequencing to evaluate the performance of different combinations of existing PCR primers to characterize orchid mycorrhizal communities associating with three orchid species (*Anacamptis morio, Ophrys tenthredinifera*, and *Serapias lingua*). The authors showed that primer pairs ITS3/ITS4OF and ITS86F/ITS4 increased the number of operational taxonomic units recovered from mycorrhizal cultures isolated from the orchids, as well as the mycorrhizal fungi present in orchid root samples compared to other primer pairs. These results highlight the importance of utilizing these primer pairs for estimating orchid mycorrhizal diversity.

9.6 CONSIDERATIONS IN UNDERTAKING A CONSERVATION GENETICS STUDY

Many researchers and conservation managers have embraced the field of conservation genetics to provide insight into various management aspects of species at risk. Yet, while results of genetic research commonly produce conservation implications for focal species at risk, they often do not translate into management outcomes (Hogbin et al. 2000). There may be uncertainty of the role conservation genetics plays in the recovery of species or perhaps a lack of communication between researchers and conservation practitioners regarding applicability of genetics (Howes et al. 2009). As molecular tools are one of a myriad of research tools available to inform recovery actions for species at risk, dialogue between practitioners and researchers is necessary to ensure that the correct molecular tool is used.

Various authors have suggested criteria for how and when molecular tools should be applied (Frankham et al. 2002; Hogbin et al. 2000; Peakall and Sydes 1996). More recently a decision framework has been proposed (Howes et al. 2009) with the decision key developed for the purpose of guiding conservation practitioners with limited (but rudimentary) understanding of conservation genetics to evaluate the practical merit of a proposed research question to species conservation (Howes et al. 2009).

9.7 SCORING MOLECULAR DATA

Genetic variation generated by the techniques described above is quantified by electrophoresis. *Electrophoresis* is the migration of particles through a gel medium under the influence of an electric current. Proteins or DNA fragments of different size migrate through the gel at different rates. Visualization of these products is by the use of stains such as ethidium bromide, or by the use of fluorescent dyes tagged to DNA when using an automated sequencer.

Co-dominant markers such as allozymes and microsatellites allow identification of heterozygotes and homozygotes at a single locus (site). Heterozygotes produce two bands on an electrophoretic gel, or two peaks, one for each allele. Homozygotes will produce a single band, often at double the intensity, as both alleles are identical. Using a variety of polymorphic loci, a wide range of characteristics within and between populations can be determined. These include: number of alleles, frequency of each allele, observed heterozygosity, and expected heterozygosity based on the Hardy-Weinberg equilibrium. This structuring of genetic variation can be used under certain assumptions to infer historical levels of gene flow, genetic drift, effective population size, levels of inbreeding, population bottlenecks, paternity analysis, and individual relatedness. Automated programs for the calculation of population genetic statistics are listed in Table 9.1.

For dominant markers, as heterozygotes cannot be distinguished from homozygotes, gels are typically scored according to the presence or absence of fragments of a particular size. To overcome this disadvantage, a large number of markers per individual are usually generated. The presence/absence data matrix can be converted to an estimate of genetic distance between individuals using a distance measure such as Euclidean distance, which can then be analyzed with clustering analysis, or by an ordination technique such as multidimensional scaling. This *genetic map* summarizes the genetic distance between all individuals analyzed. As a result, unique DNA fingerprints can be produced for genetically distinct individuals. These markers can be used for unambiguous assignment of paternity to offspring in natural populations or delineate genetic provenance for species targeted for restoration. The presence/absence data matrix is also suitable for a partitioning of variance analysis, for example, using analysis of molecular variance (AMOVA), which is analogous to a conventional analysis of variance (ANOVA).

9.8 ANALYSIS OF DATA

To understand processes driving evolution or plant rarity, the analysis of genetic diversity within and among species and populations of species is vital. A large quantity of data can now be produced at an unprecedented rate, requiring the use of dedicated computer software to extract embedded information.

Various programs are available for population phylogenetics/systematics analyses of different types of molecular markers, ranging from the computation of descriptive statistics to more specific and model-based analyses that aim to reconstruct the past demography of a set of populations, or detect loci under selection. These programs are based on many

Table 9.1 Descriptions of analytical software for population genetic analysis with example references for how they have been applied in orchid research, with an emphasis on threatened species

Name and Reference	Description	Analysis Types	References
ARLEQUIN Excoffier et al. (2005) http://cmpg.unibe.ch/software/arlequin3/	Enables the user to extract genetic and demographic features from intra- and inter-population data sets. Contains a graphical interface for users to easily select analysis types.	GDiv, HWE, LD, F_{ST}, AMOVA, population, assignment test	Forrest et al. (2004), Bhattacharyya et al. (2013); Fay et al. (2009); Pillon et al. (2007); Chen et al. (2013)
POPGENE (Yeh et al. 1997) https://sites.ualberta.ca/~fyeh/popgene.html	Analysis of genetic variation among and within populations using co-dominant and dominant markers. Windows graphical user interface with simple menus and dialogue box selections.	GDiv, GD, G statistics, F_{ST}, HWE, LD, gene flow tests	Wong and Sun (1999); Li and Ge (2006); Chung et al. (2012); Ding et al. (2008); Wallace (2002); Rodrigues and Kumar (2009)
GENEPOP (Raymond and Rousset 1995) http://genepop.curtin.edu.au/	Web-based program which analyzes haploid or diploid data. It also converts the input GENEPOP file to a variety of other formats used by other popular programs.	GDiv, Nm, HWE, LD, F_{ST} and Rho stats, Mantel tests	Swarts et al. (2009); Paul et al. (2013); Gustafsson (2003); Ross and Travers (2016); Menz et al. (2015); Pellegrino et al. (2006)
TFPGA (Miller 1997) http://www.marksgeneticsoftware.net/index.html	Windows-based program with the ability to analyze hierarchical data sets as well as data from either co-dominant markers such as allozymes or dominant markers such as AFLPs or RAPDs.	GDiv, GD, F_{ST}, HWE, Mantel tests, UPGMA cluster analysis	Li et al. (2008); Qian et al. (2014); Tali et al. (2006); Wallace (2002); Machon et al. (2003); Izawa et al. (2007)
GENALEX (Peakall and Smouse 2006) http://biology-assets.anu.edu.au/GenAlEx/Welcome.html	User-friendly cross-platform package that runs within Microsoft Excel. Analyzes co-dominant, haploid, and binary genetic data applicable to plants, animals, and microorganisms.	GDiv, GD, G statistics, F_{ST}, HWE, LD, AMOVA, Mantel tests, PCoA	Cozzolino et al. (2006); Swarts et al. (2009); Flanagan et al. (2006); Qian et al. (2013); Duffy et al. (2011); Yang et al. (2014)
STRUCTURE (Pritchard et al. 2000) http://pritchardlab.stanford.edu/structure.html	Analyzes multi-locus genotype data including SNPs, microsatellites, RFLPs, and AFLPs to investigate population structure and infer if individuals are migrants or admixed.	Allele frequencies, assigns individuals to populations, hybrid zones	Pfeifer et al. (2009); Worley et al. (2009); Bhattacharyya and Kumaria (2015); Pinheiro et al. (2010)

GDiv: Genetic diversity measures which include: number of alleles in a population (N_A), number of effective alleles (Ne), observed heterozygosity (H_O), expected heterozygosity (H_E)

GD: Genetic distance

F_{ST}: Population differentiation index (Wright 1965)

HWE: Hardy Weinberg Equilibrium

LD: Linkage disequilibrium

AMOVA: Analysis of molecular variance

UPGMA: Unweighted pair group method with arithmetic mean

PCoA: Principle coordinate analysis

Table 9.2 Descriptions of analytical software for phylogenetic analysis with example references for how they apply to orchid taxonomy and orchid mycorrhizal fungi associations, particularly threatened orchid taxa

Name and Reference	Description	Statistical Analysis	Example References (orchid/ mycorrhiza phylogenetics)
MEGA (Tamura et al. 2011) http://www.megasoftware.net/	Conducts automatic and manual sequence alignment, infers phylogenetic trees, mines web-based databases, estimates rates of molecular evolution, and tests evolutionary hypotheses.	Maximum likelihood, distance method, ordinary least squares, maximum parsimony, composite likelihood, Bayesian	Zhai et al. (2013); Luo et al. (2014); Hu et al. (2016)
			Luca et al. (2014); Oliveira et al. (2014); Xing et al. (2013)
PAUP (Swafford 1999) http://paup.sc.fsu.edu/	Software package for the inference of evolutionary trees with the ability to test for combinability of data and permutation tests for nonrandomness of data structure.	Parsimony, maximum likelihood, distance methods	Linder (1995); Pillon et al. (2006); Bateman et al. (2008); Devey et al. (2008)
			Otero et al. (2004); Martos et al. (2012); Swarts et al. (2010)
PHYLIP (Felsenstein 1989) http://www.phylip.com/	Package of programs for inferring phylogenies (evolutionary trees) from molecular sequences, gene frequencies, restriction sites and fragments, distance matrices, and discrete characters.	Parsimony, distance matrix, and likelihood methods, including bootstrapping and consensus trees	Szalanski et al. (2001); Leles et al. (2015); Campbell et al. (2007)
			Yamato et al. (2005); Pellegrino et al. (2016)
MrBayes (Ronquist and Huelsenbeck 2003) http://mrbayes.sourceforge.net/	Command line software for Bayesian inference across a wide range of phylogenetic and evolutionary models. The Markov Chain Monte Carlo (MCMC) method is used to estimate the posterior distribution of model parameters.	Bayesian analysis, convergence diagnostics, combinations of positive, negative, and backbone constraints on topologies, multi-species coalescent using a range of algorithms	Micheneau et al. (2008); Eum et al. (2011); Bytebier et al. (2008); Salazar et al. (2009)
			Jacquemyn et al. (2012); Bailarote et al. (2012); Phillips et al. (2014); Nomura et al. (2013)
PhyML (Guindon et al. 2010) http://www.atgc-montpellier.fr/phyml/binaries.php	Phylogenetic analysis software based on the maximum likelihood principle on nucleotide and amino-acid data sets of various sizes.	Maximum likelihood using Nearest Neighbor Interchanges for tree topology	Menz et al. (2015); Deng et al. (2015); Guo et al. (2012)
			Shefferson et al. (2008); Shefferson et al. (2007); Phillips et al. (2016); Pedersen et al. (2009)
RAxML (Stamatakis 2006) http://sco.h-its.org/exelixis/software.html	Randomized accelerated maximum likelihood is a sequential and parallel program for inference of large phylogenies with maximum likelihood.	Maximum likelihood and maximum parsimony, arrangement of tree topologies and branch length with descriptors	Salazar et al. (2014); Barrett and Freudenstein (2011); Freudenstein and Senyo (2008); Breitkopf et al. (2015)
			Barrett et al. (2010); Jacquemyn et al. (2011); Yuan et al. (2010); Girlanda et al. (2011); Ruibal et al. (2014); Roche et al. (2010)

documented assumptions that need to be integrated into the interpretative framework of the users. Most data analyses will require the use of more than one program, and users should start with generalist packages to uncover the basic properties of the data. Following this, specialized methodologies can be used to address more specific questions, a process that often requires many file conversions.

The broad spectrum of software available for population genetics analysis is reviewed in Excoffier and Heckel (2006) and Labate (2000). These reviews compare the value of each of the available programs, assessing their relative performance and accuracy. In addition, they describe the principles of the statistics they compute, the assumptions they make, and limitations. For phylogenetics, reviews of available software and analytical approaches can be found in Felsenstein and Felsenstein (2004), Morrison (2005), Anisimova et al. (2013), and specifically for orchids (Rasmussen 2000). In addition to these, there are dedicated reviews of more specialized programs and approaches that implement new and advanced methodologies in Bayesian analysis (Beaumont and Rannala 2004), maximum likelihood (Nielsen and Slatkin 2000), gene flow (Neigel 1997), and population differentiation (Raymond and Rousset 1995). Examples of the more frequently encountered genetic analytical software available for population genetics (Table 9.1) and phylogenetics (Table 9.2) research are described.

Case Study 9.3

Using genetic data to inform conservation management—Cypripedium calceolus

Michael F. Fay, The Royal Botanic Gardens, United Kingdom

Cypripedium calceolus (lady's slipper orchid) is widespread in temperate Eurasia (Figure 9.6), but is often rare in many countries where it occurs. In England the species was reduced to a single known wild plant in the early twentieth century as a result of excessive plant collection (Figure 9.5). In addition, there are several cultivated plants of known English origin and some with less well-documented origins that may have other origins (Fay and Taylor 2015).

Figure 9.6 *Cypripedium calceolus* (photo: M. J. M. Christenhusz)

Since the 1980s, this species has been the focus of a major conservation effort in England, involving wardening of plants, hand pollinations, *in vitro* germination of seed, and reintroduction of seedlings to locations in its former range in northern England (Ramsay and Stewart 1998). An early decision was to only introduce seedlings resulting from self-pollination of the wild plant to the wild site to avoid genetic erosion of any local adaptation. At other introduction sites, however, plants from cross-pollination between native plants (wild or cultivated) were considered to be of low risk; plants are now growing at more than a dozen sites and have flowered at several of these (Fay and Taylor 2015).

In the 1990s, Kew was requested to provide genetic support for the propagation program. Questions related to the number of genets among the wild and cultivated plants and the native (or non-native) status of the cultivated plants, with the information being used to inform the pollination program, maximizing genetic variability among the progeny and ensuring that only native genotypes were used. However, due to the large genome size of this species, genetic fingerprinting (notably AFLP; Fay et al. 2005) proved problematic, and it is only as a result of the development of NGS technology for nuclear microsatellite discovery that many questions have now been answered.

Early studies based on plastid DNA variation (Fay and Cowan 2001; Fay et al. 2009) revealed that some cultivated plants were almost certainly not native (as they were more closely related to plants from Central Europe than to the English plants of known origin) and these were excluded from the propagation program. Recently, two sets of nuclear microsatellites [based on English and Polish (Minasiewicz and Znaniecka 2014) material] have been developed and 12 loci provide effective genotypes for the plants. This has revealed several genotypes among the cultivated plants, indicating that although there has been a severe bottleneck in terms of numbers, some genetic variation has been maintained [compared to the larger, but genetically depauperate populations in Denmark (Rasmussen et al. 2012)]. However, some cultivated plants have identical genotypes and are almost certainly clonal divisions of one individual.

The information from these microsatellite studies is now being used to inform the cross-pollination program, while providing the opportunity to generate novel genotypes and create more variable introduced populations, thus minimizing the impact of the narrow gene pool available in England.

9.9 TECHNICAL NOTES: DNA EXTRACTION

Perhaps the most critical step in molecular analyses is the extraction of pure (clean), high-molecular weight DNA. The search for an efficient means of extracting DNA of both high quality and yield has resulted in the development of a variety of protocols, however, the fundamentals of DNA extraction remain similar. DNA must be purified from cellular material in a manner that prevents degradation. Because of this, even crude extraction procedures can still be adopted to prepare a sufficient amount of DNA to allow for multiple end uses. There are many techniques available and often methods within the technique will need to be optimized for a given species and tissue type. Procedure 9.1 explains two extraction techniques for plant DNA which have been tried and tested by the authors, with good results (see Figure 9.7).

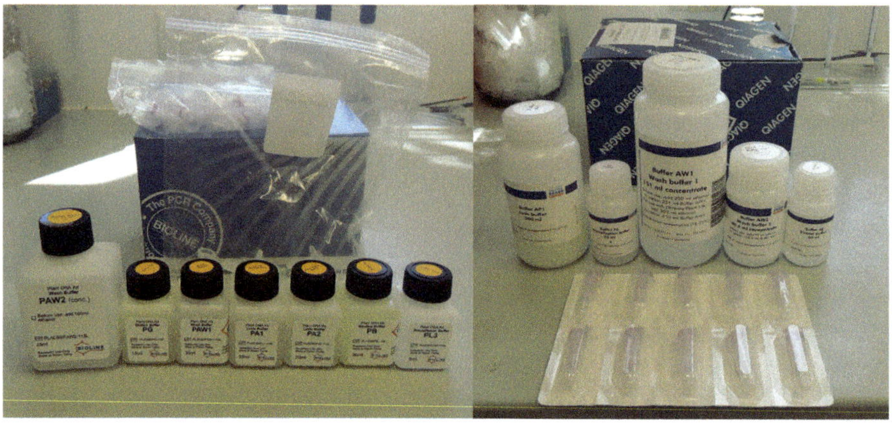

Figure 9.7 DNA extraction kits for plant or fungal material

Procedure 9.1

Plant extraction technique—DNAzol®

DNAzol is a quick, cheap, and effective procedure and works across a wide range of taxa. It is based on a novel guanidine detergent lysing solution that permits selective precipitation of DNA from a cell lysate. DNAzol® is available from Invitrogen.

1. Grind plant tissue (approximately 100 mg) in liquid nitrogen using a mortar and pestle (or micropestle and eppendorf).
2. Add 300 μl plant DNAzol and mix well.
3. Add 1–2 μl RNAse and incubate at 25°C for 1.5 hours.
4. Add 300 μl chloroform and mix vigorously. Shake for 5–15 minutes.
5. Centrifuge at 13,500 rpm for 10 minutes, transferring the supernatant to a new tube.
6. Add 225 μl cold 100% ethanol. Mix by inversion and stand for 5 minutes.
7. Centrifuge at 10,000 rpm for 4 minutes.
8. Discard supernatant. Wash pellet with 300 μl DNAzol wash (1 volume DNAzol and 0.75 volume 100% ethanol). Stand for 5 minutes.
9. Centrifuge at 10,000 rpm for 4 minutes.
10. Pour off supernatant and rewash pellet with 70% ethanol (twice).
11. Discard supernatant and dry off pellet.
12. Dissolve in 70 μl TE buffer or deionized (DI) water.

Procedure 9.2

Plant extraction technique—CTAB extraction (Doyle and Doyle 1987)

DNA may be difficult to extract from different species or different plant tissue since it may not be released from the cells. Once the tissue has been sufficiently ground, it can then be resuspended in a suitable buffer, such as CTAB, which is a detergent. When DNA has to be isolated from cells, CTAB is used to facilitate the lysis of cells to release DNA into the bulk of the solution, where it is isolated through further steps.

1. Grind leaf material in liquid nitrogen until a powder is obtained.
2. Add 500 µl CTAB extraction buffer and incubate at 65°C for 15-20 minutes.
3. Centrifuge at 11,500 rpm for 10 minutes, transfer supernatant to a new tube, and add 200 µl 5M potassium acetate (pH 4.8).
4. Mix gently and place in −20°C freezer for 10 minutes.
5. Thaw and centrifuge at 11,500 rpm for 10 minutes at 4°C.
6. Transfer supernatant to a new tube and add an equal volume of chloroform. Mix well and leave on shaker for 10 minutes at 60°C.
7. Spin at 8,000 rpm for 10 minutes, transferring top layer to a new tube. Repeat Step 5 if top layer is cloudy.
8. Add 330 µl cold isopropanol and mix gently to precipitate DNA. Place in freezer 5–30 minutes.
9. Spin at 3,000 rpm for 15 minutes to collect DNA. Remove and discard supernatant (CAREFUL: Pellet will not be solid).
10. To separate the DNA from polysaccharides dissolve pellet and viscous layer in 100 µl DI water, and then add 400 µl 2.5 M NaCl.
11. Precipitate DNA by adding 1 ml cold 95% ethanol.
12. Centrifuge at 6,000 rpm for 10 minutes, discarding the supernatant.
13. Wash pellet with 1 ml 70% ethanol (to remove salts); discard the liquid.
14. Dry and resuspend DNA in 100 µl 0.1 ml TE.
15. Incubate with 1–2 µl of RNAse enzyme at 37°C for 60 minutes.

To make 100 ml CTAB buffer

2.0 g CTAB (Hexadecyl trimethyl-ammonium bromide)
10.0 ml 1 M Tris pH 8.0
4.0 ml 0.5 M EDTA pH 8.0 (EthylenediaminetetraAcetic acid Di-sodium salt)
28.0 ml 5 M NaCl
40.0 ml DI water
1 g PVP 40 (polyvinyl pyrrolidone (vinylpyrrolidine homopolymer) MW 40,000)
Adjust all to pH 5.0 with HCl and make up to 100 ml with DI water.

In addition to the techniques described, there are many other options for orchid DNA extraction. Two commonly used methods include the QIAGEN DNeasy protocol and a method described by Jobes et al. (1995). The Qiagen DNeasy kit (cat #69104) is generally successful, fast when following the manufacturer's instructions, and does not use volatiles such as chloroform or phenol. This option is relatively expensive compared to other methods but can be used for both plant and fungal DNA extractions. Jobes' protocol offers an alternative to a CTAB procedure to efficiently remove polyphenolics, polysachharides, and RNA for cleaner DNA extraction. This technique binds the polyphenolic compounds with a high molar concentration of sodium chloride which inhibits coprecipitation of polysaccharides and DNA. This method also improves the removal of RNA by selective precipitation with lithium chloride. This is a time-consuming protocol and requires the use of highly toxic reagents (phenol, chloroform, lithium chloride), which may not be authorized in some laboratories.

9.9.1 DNA Extraction of Mycorrhizal Fungi

Successful extraction of DNA can be achieved from agar culture plates containing mycorrhizal fungi or from tubes where fungi are grown in liquid agar. Approximately 50 mg of fungal tissue is the minimum required for DNA extraction from either medium. From agar plates, fungal hyphae must be scraped off the surface of the agar with a sterile scalpel in a laminar flow cabinet to avoid contamination from airborne spores. Some fungi grow through the agar media rather than on the surface, making separation of the mycelium difficult. With liquid agar, centrifuging (Figure 9.8a) the liquid culture is helpful to separate the mycorrhizal culture from the liquid media. The liquid is removed and the solid material tipped out onto sterile filter paper. The mycelium is separated from the agar inoculation blocks with sterile forceps.

The mycelium is placed into 2 ml round-bottomed screw cap tubes with two carbide beads and snap frozen in liquid nitrogen. These can then be stored at –80°C until ready for DNA extraction with either the Qiagen DNeasy Plant Mini Kit, Bioline ISOLATE II Plant DNA Kit, or something similar. Once ready for extraction, grind the tissue in a mechanical homogenizer (Figure 9.8b) at maximum speed for up to five minutes, depending on the tissue mass, avoiding prolonged thawing of tissues. Refreeze with liquid nitrogen and repeat a few times until the tissue is well ground. Then, continue with the chosen DNA extraction kit following the manufacturer's instructions.

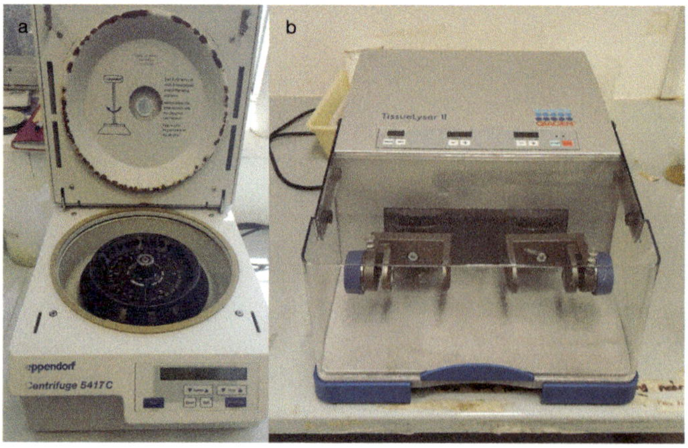

Figure 9.8 Equipment used in genetics studies (a) centrifuge and (b) mechanical tissue lyser

10

STORAGE OF ORCHID SEED AND MYCORRHIZA

10.1 BACKGROUND

While *in situ* conservation takes higher priority than *ex situ* conservation, safeguarding species through long-term storage of their germplasm, provides a necessary backup in the unfortunate event of species extinction. *Ex situ* conservation refers to *offsite* storage and maintenance of orchid seed and germplasm that is genetically representative of the species and mycorrhizal fungi shown to be germination efficacious. This conservation strategy should aim to efficiently provide material for *in situ* recovery actions (Chapter 7) such as translocations, and mitigate genetic and demographic problems associated with the production of artificial populations of plants through provenance sampling across the species geographic range.

Included under the umbrella of *ex situ* conservation is regeneration of orchids from stored seed and mycorrhizal fungi (Chapter 4) and the continued propagation of orchid species to produce conservation units. Here we detail approaches for orchid seed collection, drying, and storage; and cryostorage (ultra-low temperatures) of orchid seed and mycorrhizal fungi. Though primarily applicable to herbaceous terrestrial species, the techniques illustrated may equally apply to epiphytic species. However, techniques and approaches in seed science for orchids is an emerging scientific discipline with only a handful of laboratories worldwide researching seed-banking procedures. Thus, seed-banking of orchids requires a high level of repeated assessment of the viability of the seed. Experimentation may also be required for previously unstored orchid species where a small sub-sample of seed is tested to ensure the seed is amenable to a particular storage protocol.

10.2 SECURING COLLECTIONS OF HIGH CONSERVATION VALUE

The success of an *ex situ* conservation program depends on an ability to adequately represent the species, both during storage and recovery from storage, while ensuring the utility of stored germplasm for reintroduction programs (Husband and Campbell 2004). An outstanding opportunity for the conservation of orchids is provided by their production of dust-like seed. A single seed capsule is capable of producing many thousands of seeds with seed weights of 1–2 μg and for some species and over a million seeds have been recorded in one capsule (Arditti and Ghani 2000). Though technically very difficult to count, it is reasonable to reach the minimum accession of approximately 10,000 seeds (Way 2003). However, this can come from just one seed capsule, highlighting problems that might be associated with limited genetic diversity. It should be noted thus, that for small flowered and small capsuled orchid species, such as *Lepanthes stelis*, many more capsules are required to meet minimum accession number requirements.

Seaton and Pritchard (2003) suggest that samples of orchid seed representing all the world's orchid flora theoretically could be stored in the space equivalent to two domestic refrigerators. Other forms of germplasm that can be stored include pollinia, germinated protocorms, mycorrhizal fungi, and tubers/dormant perennating structures (tubers), however, these provide only short-term (less than one year) storage.

Long-term storage of germplasm involves the use of ultra-low temperatures (−20°C to −196°C) using liquid nitrogen (LN) to reduce the incidence of conservation artifacts such as tissue/organ aging, spontaneous (synthetic) genetic variation (potentially nonadapted variation), or the gradual breakdown of cellular integrity (Touchell and Dixon 1993). Batty et al. (2001) found that dried orchid seed stored after one year in LN germinated substantially better than freshly collected seed or seed stored at 4, 18, and 22°C for a similar period.

Deciding the species and type of material to be stored in *ex situ* collections depends upon developing priorities. The priorities for each collection program will, in turn, depend upon financial considerations, location, and resources available. Priorities for species selected for seed collections include:

- Level of immediate extinction threat/conservation status
- Level of endemism
- Provenance issues and relative distribution of the species
- Availability of seed and capacity to revisit sites
- Type of material to be stored
- Levels of fruit set
- Seasonal constraints
- Consent to collect by regional/national authorities*

10.3 SEED QUALITY

The value of a conservation collection of seed depends ultimately upon storing the highest possible quality (i.e., high viability and germinability). Maximizing seed quality at the time of collection and during transit will significantly improve long-term storage potential. Seed quality depends on interplaying factors—such as age of the flower at the time of pollination, pollen age and quality, climatic conditions during capsule ripening, parent plant vigor, levels of outcrossing or inbreeding, time of collection, and post-harvest treatment (Seaton and Pritchard 2003).

Physiological and physical changes occur in the developing embryo during the seed maturation process that can be from weeks to 10 months in some orchids. One of the most significant changes is the gradual decline in seed moisture as the embryos accumulate dry matter (seed filling phase). Just prior to dehiscence, the seed enters an abscission phase

*Where orchid germplasm is moved across international borders, it is critical that all necessary permits are in place, particularly CITES certification. Consult your local conservation office for information.

where the connection between the seed and parent plant is severed; and this phase is often the point when dormancy is instilled in the embryo.

Deterioration in seed quality may begin at any point in the plant's development stage from fertilization onward. Seed quality depends upon environmental conditions of the maternal environment, as well as seed handling during the harvesting, transport, processing, and storage phases. Ensuring the seed is maintained in a state that avoids fungal attack is important—such as maintaining the pods in a cool and dry state during transport.

Reduction in seed quality can occur from the moment of harvest, thus it is critical to begin the processing of material for long-term storage (*banking*) as soon as practicable. This includes actions such as removing the seed from capsules to prevent contamination by bacteria and fungi and drying the seed to optimal moisture content (usually at 15% relative humidity). Extra effort should be made to collect seeds at an appropriate stage of development. Wherever possible, collecting dispersal units (seeds or fruits) when they are close to the point of natural dispersal such that they tolerate rapid, enforced drying and banking is a good long-term investment.

An effective way to determine seed quality is to observe a sample of seeds under a dissecting microscope. Seed capsules will generally contain a mixture of full embryos, shrunken and dead embryos, and empty seed coats. Although the presence of plump embryos is a good indicator of high quality seed, it is still recommended that viability and—more important—germination testing (Chapter 4) is undertaken to obtain a more accurate indication of seed quality.

10.4 ORCHID SEED BANKING

Knudson (1954) first reported that dry seeds of some orchid species could be stored for at least 20 years at room temperature. The benefits of seed drying are similar between orchid and crop seeds and the principle of drying is as pertinent and valuable for orchids as for non-orchid species. Thus, it is likely that some orchid species are capable of tolerating dry storage, probably for decades, when stored at −20°C (Seaton and Pritchard 2003). Also due to their minute size and weight, large numbers of seeds can easily be stored in a small volume, making orchids ideal subjects for seed banking without the need for large facilities.

A significant outcome of the 1984 World Orchid Conference held in Miami was the proposal that the orchid community should consider banking orchid seed as insurance against possible extinction in the wild. However, despite the popularity of orchids, seed banking is in its infancy with few empirical studies and a lack of general guiding principles, unlike the sophisticated approaches developed for seed banking of non-orchids (see www.kew.org/science-conservation/collections/millennium-seed-bank). The orchid species investigated so far display a diverse array of responses to standard seed storage conditions, even though seed size and morphology is similar between orchid species. Seed responses to storage conditions vary from retaining high viability at sub-zero to −70°C, such as in *Dactylorhiza*, *Dendrobium*, *Eulophia*, and *Paphiopedilum* (Pritchard 2003; Pritchard et al. 1999), to a substantial loss of viability found in the tropical epiphyte *Cattleya aurantiaca* after seven years in storage (Seaton and Hailes 1989), or a partial loss of viability in two out of three co-occurring native Australian terrestrial orchids (Batty et al. 2001). In some orchid species, storage at sub-zero temperatures for extended periods can lead to an increase in germinability (Batty et al. 2001) and may be associated with lipid body dissociation during the freeze-thaw cycle (Pritchard 1984). Thus, prior to embarking on an orchid seed storage program, a degree of empirical research is required to establish appropriate seed moisture and temperature conditions to ensure the longevity of stored seed.

In a recent review, Merritt et al. (2014) identified inconsistencies in the storage behavior of orchid seeds at ultra-low temperatures. Although seeds from most orchid species are desiccation tolerant, this does not guarantee long-term survival at low temperatures. The same study recommended that "*In the absence of any prior knowledge, seed banks would be well advised to closely monitor the seed viability of a phylogenetically diverse representation of species during the first months of storage to screen for intermediate storage behavior*". Some species were shown to exhibit nonorthodox behavior in storage, and it was recommended that if the species was of high conservation value, seeds of these species may benefit from treatment with cryoprotectants (Merritt et al. 2014). If the seed collection of a species is of a reasonable size and freezer space is available, a useful strategy to mitigate the risks of reduced viability in ultra-low temperatures is the use of multiple storage environments (i.e., −18 to −196°C).

In many cases, botanic gardens provide the ideal venue for *ex situ* collections with their established horticultural skill, scientific capacity, and institutional support that ensures collections are scientifically managed and curated for the long term. Consideration must be given to the long-term strategic direction of botanic gardens, as often priorities shift with changes in management. This may compromise the security of long-term collections. Long-term funding options—such as through endowments or enduring Foundation support—is critical if seed banking of orchids is to be undertaken in the private or University sector.

Text Box 10.1

Orchid seed banking projects

One of the most significant seed banking initiatives is the *Millennium Seed Bank Project*, which is managed by the former Seed Conservation Department of the Royal Botanic Gardens Kew (see Figure 10.1). The major aims of this project are to conserve seed samples of the native UK

Figure 10.1 Millennium Seed Bank at Kew

flora and 25% of the world's flora by 2020. Orchids comprise a significant component of these collections with global collaboration required to ensure that targets are met. Two examples of these collaborations include: The Kenyan *Seed for Life* project and the *Orchid Seed Bank Challenge* in southwest Australia.

The Kenyan *Seed for Life* project, initiated by the National Museums of Kenya and based on an existing program in terrestrial orchid conservation, has resulted in the collection and storage of at least 25% of Kenyan orchid species, representing over half of African orchid genera.

In the southwest Australian global biodiversity hotspot (Myers et al. 2000), *The Orchid Seedbank Challenge* aims to collect and store all orchid taxa from a global biodiversity hotspot. Supported by the RBG Kew Millennium Seedbank Project, the endeavor has now achieved seed storage and mycorrhiza isolation for three quarters of the 408 native terrestrial orchid species (Brown et al. 2008) found in the southwest Australian biodiversity hotspot and is well on its way to achieving the 100% target. This includes 70 rare and critically endangered orchid taxa in addition to their mycorrhizal fungi.

The *Orchid Seed Stores and Sustainable Use* (OSSSU), aimed to establish a global network of orchid seed banks, with an initial focus on countries with high orchid diversity in Asia and Latin America. This was achieved by funding joint projects between UK institutions and another 27 partner countries with particularly active organizations in China; Singapore; Thailand in Asia; and Columbia, Ecuador, Brazil, and Bolivia in Latin America. The value of OSSSU is in capacity-building conservation-based collections of orchid seed and also generating data as species are being germinated on at least two different media for 240 taxa, representing a broad range of species and genera. Further international initiatives such as OSSSU provide a blueprint for capacity building in developing countries, and it is hoped that future initiatives of this type will emerge over time.

Case Study 10.1

*Botanic gardens in orchid conservation—*ex situ *seed storage at the Atlanta Botanical Gardens*

Jennifer Cruse-Sanders
Atlanta Botanical Garden, USA

The epicenter of North America's orchid diversity is found at the southern end of Florida in the wild habitat of the Fakahatchee Strand Preserve State Park with 48 species of native epiphytic and terrestrial orchids. With funding from the Association of Zoological Horticulture, the Atlanta Botanical Garden has partnered with the Florida Park Service, Florida International University, and the Florida Fish and Wildlife Conservation Commission to safeguard six species of rare orchids from the park including: *Dendrophylax lindenii, Encyclia pygmaea, Pleurothallis gellida, Epidendrum strobiliferum*, and *Epidendrum floridense*. Surveys were conducted during flowering and fruiting seasons; and seeds were collected and brought back to the Garden for long-term storage (Figure 10.2).

After a percentage of the seeds were stored in the freezer or initiated in the micropropagation laboratory, excess seed was returned to the park to be sown into appropriate habitats.

Figure 10.2 (a) *Cyrtopodium punctatum*; (b) *Encyclia pygmaea*; and (c) *Pleurothallis gellida* (photos: Jay Staton).

Another long-term partnership for orchid conservation within the Fakahatchee Strand is restoration of the flagship epiphytic orchid, the cigar orchid, *Cytropodium punctatum*. By 2008, its population in the Fakahatchee Strand was reduced to only 23 plants as a result of poaching and past clear-cut logging. Efforts to restore this orchid from 2008 to 2015 included 34 manual pollinations of wild plants with 14 fruits collected for propagation: 10% of the seed stored, 10% propagated, and 80% returned to the natural population.

The *ex situ* propagation program at the Atlanta Botanical Garden has produced large numbers of orchids for restoration with 1,202 cigar orchids reestablished into 18 locations in the Park with 60% survival. Follow up investigations will involve monitoring orchid survival and estimates of population viability.

10.5 VOUCHERING SPECIMENS

Germplasm collections have little value in the context of conservation if the material is not adequately identified, vouchers taken and recorded, and storage material adequately curated (Schuiteman and de Vogel 2003). In the process of germplasm collection and storage, accurate identification of both orchid and mycorrhizal fungus should be coupled with an understanding of the reproductive biology of the species to be conserved. These studies are assisted by information gathered from scientific research and publications, local botanists, herbarium records, existing seed collections, threatened status/species lists, and hobbyists in local community groups.

Herbarium vouchers are important for new accessions or where a species is suspected as representing an ecologically distinct population. Vouchers comprise dried plant material consisting of pressed leaves, stems, flowers, fruits and/or roots, preserved plants in spirit, or, in the case of rare species, even a photograph (Figure 10.3). Vouchers with attendant locality and location information are important to conservation as they provide the physical evidence to confirm the presence of plant species at specific locations at a certain time (refer to Table 2.1 in Chapter 2 for an example collection record). Voucher specimens should be

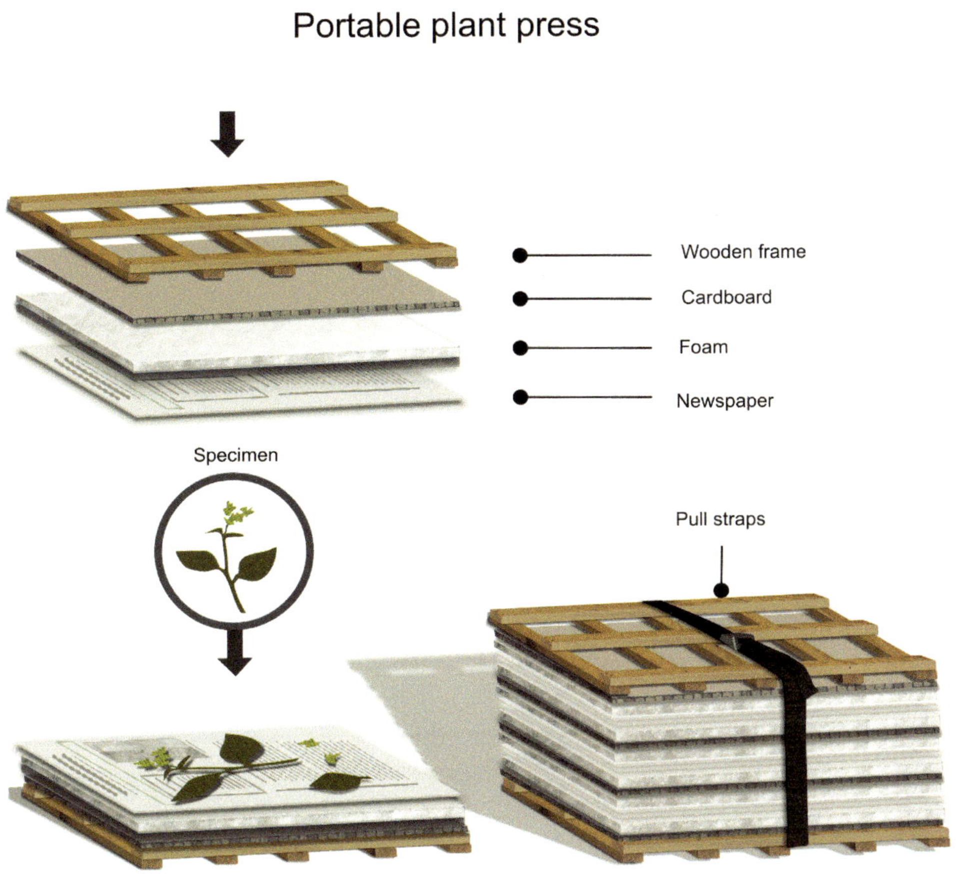

Figure 10.3 Process for vouchering plant specimens in a portable plant press

lodged with an accredited herbarium or facility mandated for long-term curation of herbarium specimens.

A photographic voucher is an image that shows sufficient detail of plant characteristics to allow identification by relevant experts. While herbaria generally prefer actual plant specimens, a photograph can be useful where specimen collection is not possible. Photographs can also accompany a dried or fresh plant specimen, to help with identification and to show the habits of plants in the field. As with all vouchers, relevant data must be maintained for each plant photographed. Macro-photographs of flowers including inflorescences, fruits, seed heads, leaves (upper and lower surfaces), buds, roots, and particularly diagnostic features such as calli or labellum detail are useful for creating a diagnostic photographic voucher. Photos with a scale should be taken of the whole plant and within the field context.

10.6 GENETIC CONSIDERATIONS IN SEED COLLECTION PROGRAMS

The selection and collection of genetically representative germplasm is an important part of the long-term aim of *ex situ* conservation programs with the goal of the creation of self-sustaining re-wilded populations of orchids. Advances in molecular methods (Chapter 9) have increased the technical capacity to target genetic provenances considered significant for conservation. Indeed, the restoration of species and habitats call explicitly for the consideration of genetic issues when plant material is prepared for translocation (Chapter 7), mixed as recommended in recovery plan implementation (Chapter 2), or stored in long-term, off-site gene banks (Fay and Krauss 2003; Rossetto et al. 1995). Appropriate genetic screening and careful plant selection is necessary to reduce the likelihood of inbreeding or outbreeding depression and to facilitate the progression of evolutionary processes post reintroduction/translocation.

As many orchid capsules can contain the recommended minimum accession of 10,000 seeds, there is a risk of reduced genetic representation if seed from only one capsule is stored. It is highly recommended that when an accession is made, that seeds from multiple plants within a population are used for provenance representation.

Genetic drift has been identified as a risk to the long-term storage of seed in seed banks (Basra 2006). Genetic drift occurs when there is a change in allele frequency from the parent population that could arise during storage. This may have an impact on the genetic makeup and survival of regenerating seedlings from that collection. Although possible, the negative effects of poor field sampling—such as limited outcrossing or collecting under-ripe seed (Chapter 4) and poor regeneration practices (Chapter 7)—is likely to be more significant than the potential for genetic drift to occur in storage. This is a research area requiring further attention, particularly as seed collections are becoming an important tool for orchid conservation programs.

10.7 LIVING COLLECTIONS AS AN *EX SITU* CONSERVATION TOOL

Living collections are an alternative *ex situ* method often used by botanic gardens as a supplement to seed banking (Figure 10.4). While botanic gardens generally contain samples of horticultural taxa (Linington and Pritchard 2001), there has been a recent focus on growing difficult-to-grow orchid taxa or specifically managing orchid collections for conservation and translocation purposes (Havens et al. 2006; Maunder et al. 2001). Plants growing in living collections in botanic gardens are an important resource for educational and research purposes, but cultivation specifically for conservation is resource intensive (Ramsay and Dixon 2003), and a number of species have been lost from living collections through poor horticultural practice or oversights in curation focus and management (Figure 10.5).

Conservation-based living collections need to be curated in a way that adequately represents the genetic diversity of the species—this means care to ensure genetic representativeness is maintained during curation and propagation cycles (Nash et al. 2003). Botanic gardens (as well as amateur/hobbyist growers) tend to have either one specimen, or a very small number of clones of any one species. If inadvertent, genetic drift has occurred during repeated propagation cycles, such propagated plants may not survive the rigors of their natural habitat. Thus, a living collection should aim to grow as wide a gene pool as possible, minimize repropagation cycles (especially if seed-based), and employ genetic screening to maximize diversity in the collection (Seaton et al. 2010).

Maintaining only a living collection of a species—particularly one that is genetically representative—for the long term can present curation challenges. Even within the

Figure 10.4 Living collections in botanic gardens provide an ongoing source of seed and propagated material for translocation programs. Conservation collection of *Paphiopedilum javanicum*—the only slipper orchid native to Bali and close to extinction on the island due to illegal collection for the nursery trade.

Figure 10.5 Some of the world's rarest orchids command high prices, with artificial propagation from nursery-generated seed helping to relieve the illegal collection of wild plants; seen here is an artificial propagation of *Paphiopedilum rothschildianum* where single shoots sell for US$6,000.

best-maintained collections, plants have a limited lifespan, often being subject to aging, disease, and unintentional loss. Plants may be lost due to heating failures in temperate climates (or indeed cooling failures in warm climates); pests or disease; or poor cultural practices. In order to minimize such losses, there is a strong case to be made for increasing support for botanic gardens in the countries of origin—and within each country to establish botanic gardens at strategic locales (e.g., altitudinally) to accommodate plant cultural requirements without the need for artificial environments. Most important, institutional support is critical to ensure that human and financial support are available for adequate, long-term maintenance of a living collection.

More recently, the need to involve the wider orchid growing community in orchid conservation through cultivation has been recognized. Members of orchid societies and other community groups have the potential to make an important contribution to orchid conservation. Orchid growers and the commercial orchid growing sector are a largely untapped pool of expertise in growing and propagating orchid taxa—often developing effective cultivation techniques long before botanic gardens and scientists. Growers are also adept in adapting techniques and technologies for successful propagation of difficult genera, such as the outstanding successes achieved in the private sector for industrial-scale production of the *Cypripedium* species (see issues of the Orchid Review, American Orchid Society, and Journal of the Hardy Orchid Society). Accessing the grower knowledge base can provide a powerful tool in overcoming cultivation difficulties and could substantially add to the conservation of species (Nash et al. 2003).

Procedure 10.1

Method for the long-term storage of terrestrial orchid seed

The methods described here for the storage of orchid seed are based on protocols recommended by the Orchid Seed Stores for Sustainable Use project (OSSSU).

1. Aim to obtain a good sample size (20,000 seeds or more than five capsules). The number of seed capsules required to collect this many seeds varies according to the species. Precise counting of orchid seed is not practical given their small size, however, estimates of total seed count can be made by weighing three replicates of up to 200 seeds on a five decimal place balance. Weights are averaged and then divided into the total weight of collected seed to obtain a total seed count.
2. Pool the samples (seed pods) of the same species or provenances and mix thoroughly (this will assist in maintaining the maximum genetic diversity in a propagation cycle).
3. Dry/equilibrate seed over a saturated solution of lithium chloride (LiCl). The relative humidity (RH) will be approximately 12% at 20°C. Equilibrating seed can also be done in a dedicated seed drying room, which will usually be at 15% RH and 15°C.
4. To make a saturated solution of LiCl, add and dissolve the salt in warm water until excess crystals remain in the bottom of the vessel.
5. Bring the solution to a gentle boil to dissolve the crystals. On cooling, some undissolved salt should remain.

6. Place seeds as a thin layer in a desiccator (seal with Vaseline) for seven days at approximately 15°C.
7. Transfer seeds (to avoid change in seed moisture content) to hermetically-sealed tubes. Tubes should be as full of seed as possible so that the seed occupies much of the air space.
8. Small vials of seed can be stored within larger tubes, within a larger, air-tight storage jar.
9. Individual storage tubes and their containers should all be labeled.
10. Storage vessels should contain sachets of silica gel with a color indicator to demonstrate that seals have not been broken and moist air has not entered the storage vessel. Prior to use, the sachets of silica gel should be equilibrated over the saturated LiCl solution. Orange silica gel begins to turn green when the RH approaches 22%.
11. Store the tubes of seed in a freezer at around −20°C or place them in LN.

10.8 LONG-TERM STORAGE OF ORCHID-MYCORRHIZAL FUNGI

For orchids, securing high value collections often requires use of a germination-compatible mycorrhiza (Chapter 4). While orchid seed can be germinated via asymbiotic *in vitro* methods, the establishment of terrestrial orchid seedlings to field sites is arguably more successful (Reiter et al. 2016) when seeds have been germinated with compatible mycorrhizal fungi (Chapter 7), although as shown in Chapter 4, some asymbiotically germinated terrestrial taxa such as *Cypripedium* and all epiphytic taxa collect mycorrhizal fungi from growth medium and substrates.

The mycorrhizal collection process must involve germination assays investigating compatibility and efficacy of mycorrhizal fungi. More important, a germination enhancing fungus may be very different to the fungus required for adult plant growth as reported for the heteromycotrophic species *Gastrodia* (Xu and Guo 2000). Such an approach also provides the capacity to test for changes in efficacy and germination rates over time. Mycorrhizal fungi can be stored using a variety of methods from temporary storage on agar slopes in tubes to lyophilization and freezing (Gams 2002), however, little remains known of the long-term effects of storage conditions on mycorrhizal viability and efficacy.

There are several problems associated with the long-term storage of fungal symbionts. Studies have shown that fungal cultures stored on agar at room temperature or at 4°C can lose viability and may require subculturing on a regular basis (Smith and Onions 1994; Zettler et al. 2003). The subculturing process can be time consuming and may lead to accidental contamination, gradual changes in culture morphology, and changes in germination efficacy (Dixon 1987). Other methods are available for long-term conservation, including storage under sterile water or mineral oil. Using these methods, cultures have been found to retain their viability for up to 20 years (de Capriles et al. 1989; Smith and Onions 1994), however, response to these storage approaches can depend upon the fungus involved (Smith and Onions 1994) with few long-term studies of orchid mycorrhiza storage.

Cryopreservation (storage in LN) is considered the most effective for retention of viability (Chapter 4) and maximizing the genetic integrity of the stored mycorrhizal fungus (see

Procedure 10.2). Batty et al. (2001) demonstrated that mycorrhizal fungi isolated from four genera of Australian terrestrial orchids could be stored in LN with no significant deleterious effects on fungal regeneration.

Case Study 10.2

The North American Orchid Conservation program—a model for global orchid conservation?

Dennis F. Whigham, Founding Director of the North American Orchid Conservation Center (NAOCC), Smithsonian Environmental Research Center

The Smithsonian Institution and the U.S. Botanic Garden joined forces to establish the NAOCC. The goal of the NAOCC is to assure the survival of native orchids in the U.S. and Canada, but the ultimate goal is to develop a model system that can be applied to global orchid conservation. The NAOCC model is based on ecological principles, but to be effective and successful it must reach beyond basic and applied research to include community and citizen scientists. Success will also depend on the development of a network of collaborating public and private organizations and individuals to act as knowledge hubs and provide *ex situ* infrastructure.

The NAOCC model is based on four guiding principles. First, seeds of native orchids will be collected from across their range of distribution to assure genetic diversity is represented in local, regional, and national seed banks. Second, because orchids are dependent on fungi at critical life-history stages, orchid conservation must also include the collection, identification, and culturing of representative orchid mycorrhiza in regional and national fungal banks. Similar to seeds, the fungi bank should be representative of the genetic diversity of the fungi. NAOCC collaborators, especially botanical gardens, are integral to the third element of the conservation model.

Botanical gardens and competent organizations, societies, and individuals will collaborate with the NAOCC to develop protocols to propagate native orchids with the goal of learning how to establish sustainable populations in cultural environments. The sustainable populations will be composed of genetically appropriate material (orchids and fungi) from each region. Once protocols have been established to develop mini-ecosystems that support orchids and fungi, citizen scientists will also become actively engaged in conservation efforts by using the species protocols to put species back into the landscape to assure that seeds and fungi are available to support the movement and establishment of populations. The fourth element of the NAOCC model is education. We live in an era when people have less and less connection to nature. Using orchids as model systems, the NAOCC will strive to develop educational activities that support an expansion of botanical literacy as well as engage citizen scientists of all ages in conservation.

The NAOCC is steadily growing with a focus on the development of a network of collaborators to support the core elements of the conservation model. To date, more than 40 organizations and many individuals have joined the NAOCC network, and regional groups have been formed across the U.S. to initiate efforts to collect and store orchid seeds and fungi. Educational and outreach efforts have also started, such as the development of 25 paper punch-out *orchid-gami* models shown in Figure 10.6.

The NAOCC web site (http://northamericanorchidcenter.org/) provides news as well as informative videos about the NAOCC and an interactive website (Go Orchids) that includes

Figure 10.6 *Orchid-gami* models used in public education and orchid conservation promotion programs by the NAOCC program of the Smithsonian Institute in the U.S. Four of 25 native *orchid-gami* models are illustrated. You can visit the species page for each orchid on the NAOCC Go Orchids website. To obtain pdf files of these and the other models, contact oneillj@si.edu—once you download the pdf files, all you need is a two-sided color printer, scissors, and (for some models) glue.

all of the native orchids in the U.S. and Canada (http://goorchids.northamericanorchidcenter.org/). Go Orchids, modeled after Go Botany, was developed with publicly available software and was designed to be a model that could be adapted anywhere in the world. The NAOCC website has also been a successful portal for the magnificent photography of amateurs who cherish orchids.

In addition to an ever-expanding network of collaborators, the NAOCC is now reaching beyond North America. A recently funded collaboration with the U.S. Forest Service will take the NAOCC to some Pacific Islands, starting with Palau, and a European consortium is seeking funding to develop an orchid conservation effort that will focus on the training of Ph.D. students. With luck and support, the NAOCC will hopefully spur orchid conservation efforts across the globe.

Procedure 10.2

Method for the cryostorage of orchid seed and mycorrhizal fungus

Place the seed and fungus in LN dewars. Note, this can be done separately for seed and/or mycorrhizal fungus.

1. Seed: Collect the seed according to protocols in Chapter 4, and prepare for storage according to protocols highlighted earlier in this chapter; then place into 2 ml cryo-vials.
2. Fungi: Select actively growing cultures growing on quarter strength potato dextrose agar medium, and under sterile conditions subculture 6–10, 5 mm^2 blocks into 2 ml cryo-vials. **Note:** Filter-sterilized cryoprotectants, that can be added to the vial though experimentation, have shown that agar blocks with fungus survive in the absence of any cryoprotectants. If required, the following cryoprotectants have been found to provide protective capabilities: 10% dimethylsulphoxide, 10% glycerol, or a 10% glucose solution (Batty et al 2001). As many fungi species have not been tested, some experimentation may be required, particularly for Basidiomycota fungal isolates.
3. Ensure caps are fastened tightly.
4. Label each cryo-vial with appropriate coding and record the position of vials in the dewar.
5. Open the LN dewar carefully using dedicated LN safe gloves and a face mask. Note: As all LN dewars are different, please observe safety protocols recommended by the manufacturer.
6. Place cryo-vials into storage racks and plunge into the LN or store in the vapor phase if vials are not liquid tight.

To remove seed and/or mycorrhizal fungus from LN.

1. Remove cryo-vials from storage racks. This needs to be done as quickly as possible to prevent thawing.
2. Rapidly thaw seed and/or fungal cryo-vials in a 40°C water bath.
3. Sow seed using suitable methods (see Chapter 4).
4. Plate out agar blocks onto fungi isolating medium (see Chapter 3) and monitor cultures for signs of growth or contamination. Note: Thawed cryopreserved fungi can take longer to recover and commence growth.

10.9 ALGINATE ENCAPSULATION OF ORCHID SEED AND FUNGUS

This method involves encapsulation of orchid seed and fungal symbiont in a sodium alginate bead, followed by pretreatment of the bead in sucrose solution, then desiccation in a flow of sterile air. This simplifies the procedures associated with maintenance of individual collections of orchid seed and mycorrhiza. The method also enables effective long-term storage at temperatures above −196°C, without the need for frequent subculturing of fungal isolates (Sommerville et al. 2008).

The encapsulated seed and fungi could potentially be used for direct seeding at sites targeted for reintroduction or population augmentation. At present, the production of orchid seedlings for transplanting requires the establishment of fungal cultures, symbiotic germination of seed, culture of seedlings under controlled humidity, then acclimation to outdoor conditions before the transfer to a potting mix (Batty et al. 2006) (see Chapter 5). If encapsulated seed could be germinated successfully directly in the field or potting mix, the production of orchid seedlings would be greatly simplified.

The encapsulation–dehydration technique has previously been used for the cryopreservation of germplasm from a variety of plant species. Engelmann (2004) and Block (2003) reported plant material stored using these methods had high survival rates, generally accompanied by rapid and direct growth from the beads. Flachsland et al. (2006) successfully applied simultaneous encapsulation to the cryopreservation of epiphytic orchid seed and protocorms and Wood et al. (2000) successfully stored terrestrial orchid seed and associated fungal symbionts for short periods. Sommerville et al. (2008) used this method for the simultaneous encapsulation of two Australian terrestrial orchids, *Pterostylis saxicola* and *Diuris arenaria,* finding that storage for 6 months at −18°C and −196°C was most effective, with no significant impact upon the fungal recovery and germination performance of either species. However, these alginate beads performed poorly when directly transferred to potting mix (Sommerville et al. 2008).

Procedure 10.3

Method for encapsulation–dehydration of orchid seed and fungus (based on Sommerville et al. (2008))

1. Sterilize the orchid seed for five minutes in a dilute bleach solution (0.5% active chlorine). See Chapter 4 for orchid seed collection protocols.
2. Rinse sterilized seed three times with sterile deionized water.
3. Add seed to a sterile solution of 2% w/v sodium alginate.
4. Scrape the fungal hyphae from the surface of several oatmeal agar plates (see Chapter 3 for fungal subculturing methods).
5. Add the hyphae to the seed–sodium alginate mixture.
6. Mix the seed, hyphae, and sodium alginate thoroughly on a magnetic stirrer for 1 hour at 600 rpm.
7. Pipette the resulting suspension drop-by-drop into a sterile solution of 100 mM calcium chloride to form individual beads with a volume of 42.5 ± 0.3 ml.
8. Leave the beads in the $CaCl_2$ solution to harden for 20 minutes, then strain and add to a solution of 0.75 M sucrose.
9. Shake the beads in the sucrose solution for 24 hours on an orbital shaker at low speed (~115 rpm), then strain.
10. Dry the beads for 18 hours on sterile absorbent paper in a laminar flow cabinet (~24°C and 32% RH).
11. Transfer the dried beads to sterile cryopreservation vials for storage.
12. It is important to determine the moisture content of the beads. This can be done by drying three subsamples of 10 beads at 103°C for 17.5 hours.

11

REFERENCES

Ackerman JD. (1989) Geographic and seasonal variation in fragrance choices and preferences of male euglossine bees. *Biotropica* **21**(4), 340–347.

Ackerman JD, Ward S. (1999) Genetic variation in a widespread, epiphytic orchid: Where is the evolutionary potential? *Systematic Botany* **24**(2), 282–291.

Ackerman JD, Zimmerman JK. Bottlenecks in the life histories of orchids: resources, pollination, population structure, and seedling establishment. In *Proceedings of the 14th World Orchid Conference*, 1994, London, UK (ed. A Pridgeon), pp. 125–129.

Alexander C, Hadley G. (1983) Variation in symbiotic activity of *Rhizoctonia* isolates from *Goodyera repens* mycorrhizas. *Transactions of the British Mycological Society* **80**, 99–106.

Alexander MP. (1980) A versatile stain for pollen, fungi, yeast and bacteria. *Stain Technology* **55**, 13–18.

Anderson AB. (1991) Symbiotic and asymbiotic germination and growth of *Spiranthes magnicamporum* (Orchidaceae). *Lindleyana* **6**(4), 183–186.

Anisimova M, Liberles DA, Philippe H, Provan J, Pupko T, von Haeseler A. (2013) State-of-the-art methodologies dictate new standards for phylogenetic analysis. *BMC Evolutionary Biology* **13**(1), 161.

Arditti J. (1992) *Fundamentals of orchid biology*. John Wiley & Sons, Inc. New York, USA, **691**.

Arditti J. (2009) *Micropropagation of orchids*. John Wiley & Sons, Inc., New York, USA.

Arditti J, Ghani AKA. (2000) Tansley review No. 110—Numerical and physical properties of orchid seeds and their biological implications. *New Phytologist* **145**(3), 367–421.

Ayasse M, Schiestl FP, Paulus HF, Ibarra F, Francke W. (2003) Pollinator attraction in a sexually deceptive orchid by means of unconventional chemicals. *Proceedings of the Royal Society of London B Biological Sciences* **270**(1514), 517–522.

Ayasse M, Schiestl FP, Paulus HF, Lofstedt C, Hansson B, Ibarra F, Francke W. (2000) Evolution of reproductive strategies in the sexually deceptive orchid *Ophrys sphegodes*: How does flower-specific variation of odor signals influence reproductive success? *Evolution* **54**(6), 1995–2006.

Bailarote BC, Lievens B, Jacquemyn H. (2012) Does mycorrhizal specificity affect orchid decline and rarity? *American Journal of Botany* **99**(10), 1655–1665.

Baker KF, Roistacher CN. (1957) Principles of heat treatment of soil. In *The U. C. system for producing healthy container-grown plants, Extension service manual 23* (ed. KF Baker) pp. 138–161. California Agricultural Experimental Station: USA.

Barrett CF, Freudenstein JV. (2011) An integrative approach to delimiting species in a rare but widespread mycoheterotrophic orchid. *Molecular Ecology* **20**(13), 2771–2786.

Barrett CF, Freudenstein JV, Taylor DL, Koljalg U. (2010) Rangewide analysis of fungal associations in the fully mycoheterotrophic *Corallorhiza striata* complex (Orchidaceae) reveals extreme specificity on ectomycorrhizal *Tomentella* (Thelephoraceae) across North America. *American Journal of Botany* **91**, 628–643.

Barton NH, Keightley PD. (2002) Understanding quantitative genetic variation. *Nature Reviews Genetics* **3**(1), 11–21.

Baskin CC, Baskin JM. (1998) *Seeds ecology, biogeography, and evolution of dormancy and germination.* Academic Press: San Diego, USA.

Basra AS. (2006) *Handbook of seed science and technology.* Food Products Press: New York, USA.

Bateman RM, Smith RJ, Fay MF. (2008) Morphometric and population genetic analyses elucidate the origin, evolutionary significance and conservation implications of *Orchis xangusticruris* (*O.purpurea* x *O. simia*), a hybrid orchid new to Britain. *Botanical Journal of the Linnean Society* **157**(4), 687–711.

Batty AL, Brundrett MC. (2002) *Orchid Conservation Techniques Manual.* Plant Science, Kings Park and Botanical Gardens, Perth, Australia.

Batty AL, Brundrett MC, Dixon KW, Sivasithamparam K. (2006) *In situ* symbiotic seed germination and propagation of terrestrial orchid seedlings for establishment at field sites. *Australian Journal of Botany* **54**(4), 375–381.

Batty AL, Brundrett MC, Dixon KW, Sivasithamparam K. (2006) New methods to improve symbiotic propagation of temperate terrestrial orchid seedlings from axenic culture to soil. *Australian Journal of Botany* **54**(4), 367–374.

Batty AL, Dixon KW, Brundrett MC, Sivasithamparam K. (2001) Constraints to symbiotic germination of terrestrial orchid seed in a mediterranean bushland. *New Phytologist* **152**(3), 511–520.

Batty AL, Dixon KW, Brundrett MC, Sivasithamparam K. (2001) Long-term storage of mycorrhizal fungi and seed as a tool for the conservation of endangered Western Australian terrestrial orchids. *Australian Journal of Botany* **49**(5), 619–628.

Batty AL, Dixon KW, Brundrett MC, Sivasithamparam K. (2002) Orchid conservation and mycorrhizal associations. In *Microorganisms in plant conservation and biodiversity* (eds. KW Dixon and RL Barrett) pp. 195–226. Kluwer Academic Publishers: Dordrecht, The Netherlands.

Batty AL, Dixon KW, Sivasithamparam K. (2000) Soil seed bank dynamics of terrestrial orchids. *Lindleyana* **15**(4), 227–236.

Beaumont MA, Rannala B. (2004) The bayesian revolution in genetics. *Nature Reviews Genetics* **5**(4), 251–261.

Bevege DI. (1968) A rapid technique for clearing tannins and staining intact roots for detection of mycorrhizas caused by *Endogone* spp. and some records of infection in Australasian plants. *Transactions of the British Mycological Society* **51**, 808–810.

Bhattacharyya P, Kumaria S. (2015) Molecular characterization of *Dendrobium nobile* Lindl., an endangered medicinal orchid, based on randomly amplified polymorphic DNA. *Plant Systematics and Evolution* **301**(1), 201–210.

Bhattacharyya P, Kumaria S, Kumar S, Tandon P. (2013) Start Codon Targeted (SCoT) marker reveals genetic diversity of *Dendrobium nobile* Lindl., an endangered medicinal orchid species. *Gene* **529**(1), 21–26.

Bidartondo MI, Burghardt B, Gebauer G, Bruns TD, Read DJ. (2004) Changing partners in the dark: isotopic and molecular evidence of ectomycorrhizal liaisons between forest orchids and trees. *Proceedings of the Royal Society of London B: Biological Sciences* **271**(1550), 1799–1806.

Blanco MA, Barboza G. (2005) Pseudocopulatory pollination in *Lepanthes* (Orchidaceae: Pleurothallidinae) by fungus gnats. *Annals of Botany* **95**(5), 763.

Block W. (2003) Water status and thermal analysis of alginate beads used in cryopreservation of plant germplasm. *Cryobiology* **47**(1), 59–72.

Bohman B, Flematti GR, Barrow RA, Pichersky E, Peakall R. (2016) Pollination by sexual deception—it takes chemistry to work. *Current Opinion in Plant Biology* **32**, 37–46.

Bonnardeaux Y, Brundrett M, Batty A, Dixon K, Koch J, Sivasithamparam K. (2007) Diversity of mycorrhizal fungi of terrestrial orchids: compatibility webs, brief encounters, lasting relationships and alien invasions. *Mycological Research* **111**, 51–61.

Borg-Karlson AK. (1990) Chemical and ethological studies of pollination in the genus *Ophrys* (Orchidaceae). *Phytochemistry* **29**(5), 1359–1387.

Bougoure JJ, Bougoure DS, Cainey JWG, Dearnaley JDW. (2005) ITS–RFLP and sequence analysis of endophytes from *Acianthus, Caladenia* and *Pterostylis* (Orchidaceae) in southeastern Queensland. *Mycological Research* **4**(109), 452–460.

Bouzat JL, Johnson JA, Toepfer JE, Simpson SA, Esker TL, Westemeier RL. (2009) Beyond the beneficial effects of translocations as an effective tool for the genetic restoration of isolated populations. *Conservation Genetics* **10**(1), 191–201.

Bower CC. (1996) Demonstration of pollinator-mediated reproductive isolation in sexually deceptive species of *Chiloglottis* (Orchidaceae: Caladeniinae). *Australian Journal of Botany* **44**(1), 15–33.

Bower CC. (2006) Specific pollinators reveal a cryptic taxon in the bird orchid, *Chiloglottis valida* sensu lato (Orchidaceae) in south-eastern Australia. *Australian Journal of Botany* **54**(1), 53–64.

Bower CC, Brown GR. (2009) Pollinator specificity, cryptic species and geographical patterns in pollinator responses to sexually deceptive orchids in the genus *Chiloglottis*: the *Chiloglottis gunnii* complex. *Australian Journal of Botany* **57**, 37–55.

Breitkopf H, Onstein RE, Cafasso D, Schlüter PM, Cozzolino S. (2015) Multiple shifts to different pollinators fuelled rapid diversification in sexually deceptive *Ophrys* orchids. *New Phytologist* **207**(2), 377–389.

Brown A, Dixon K, Hopper S, Dundas P. (2008) *Orchids of Western Australia*. University of Western Australia Press: Perth, Australia.

Brundrett MC. (2002) Coevolution of roots and mycorrhizas of land plants. *New Phytologist* **154**(2), 275–304.

Brundrett MC. (2004) Diversity and classification of mycorrhizal associations. *Biological Reviews* **79**, 473–495.

Brundrett MC, Abbott L. (1995) Mycorrhizal fungus propagules in the jarrah forest. II. Spatial variability in inoculum levels. *New Phytologist* **131**(4), 461–469.

Brundrett MC, Bougher N, Dell B, Grove T, Malajczuk N. (1996) *Working with Mycorrhizas in Forestry and Agriculture*. Pirie Printers: Canberra, Australia.

Brundrett MC, Juniper S. (1995) Non-destructive assessment of germination and single-spore isolation of VAM fungi. *Soil Biology and Biochemistry* **27**, 85–91.

Brundrett MC, Piche Y, Peterson RL. (1984) A new method for observing the morphology of vesicular-arbuscular mycorrhizae. *Canadian Journal of Botany* **62**, 2128–2134.
Brundrett MC, Scade A, Batty AL, Dixon KW, Sivasithamparam K. (2003) Development of *in situ* and *ex situ* seed baiting techniques to detect mycorrhizal fungi from terrestrial orchid habitats. *Mycological Research* **107**, 1210–1220.
Brzosko E, Ratkiewicz M, Wroblewska ADA. (2002) Allozyme differentiation and genetic structure of the Lady's slipper (*Cypripedium calceolus*) island populations in north-east Poland. *Botanical Journal of the Linnean Society* **138**(4), 433–440.
Burgeff H. (1909) Die Wurzelpilze der Orchidaceen, ihre Kultur und ihr Leben in der Pflanzen. [The root fungi of orchids, their culture and their life in plants.]. *G. Fischer, Jena* **220**.
Burgeff H. (1959) Mycorrhiza of orchids. In *The Orchids a Scientific Study* (ed. CL Withner) pp. 361–395. The Roland Press Company: New York.
Bytebier B, Van der Niet T, Bellstedt DU, Linder HP. (2008) The phylogenetic position of the enigmatic orchid genus *Pachites*. *South African Journal of Botany* **74**(2), 306–312.
Calvo RN, Horvitz CC. (1990) Pollinator limitation, cost of reproduction, and fitness in plants: a transition-matrix demographic approach. *American Naturalist*, 499–516.
Campbell VV, Rowe G, Beebee TJC, Hutchings MJ. (2007) Genetic differentiation amongst fragrant orchids (*Gymnadenia conopsea* s.l.) in the British Isles. *Botanical Journal of the Linnean Society* **155**, 349–360.
Caswell H. (2001) *Matrix population models: Construction, analysis and intepretation*. 2nd ed. Sinauer Associates, Inc. Publishers: Sunderland, Massachusetts, USA.
Charlesworth D, Charlesworth B. (1987) Inbreeding depression and its evolutionary consequences. *Annual Review of Ecology and Systematics* **18**(1), 237–268.
Chase MW, Cameron KM, Barrett RL, Freudenstein JV. (2003) DNA data and Orchidaceae systematics: a new phylogenetic classification. In *Orchid Conservation* (eds. KW Dixon, SP Kell, RL Barrett and PJ Cribb) pp. 69–90. Natural History Publications: Kota Kinabalu, Sabah, Malaysia.
Chen XH, Guan JJ, Ding R, Zhang Q, Ling XZ, Qu B, Zhang LJ. (2013) Conservation genetics of the endangered terrestrial orchid *Liparis japonica* in Northeast China based on AFLP markers. *Plant Systematics and Evolution* **299**(4), 691–698.
Chittka L. (1992) The colour hexagon: a chromaticity diagram based on photoreceptor excitations as a generalized representation of colour opponency. *Journal of Comparative Physiology A* **170**(5), 533–543.
Chung MY, Chung MG. (1999) Allozyme diversity and population structure in Korean populations of *Cymbidium goeringii* (Orchidaceae). *Journal of Plant Research* **112**(1106), 139–144.
Chung MY, Lopez-Pujol J, Maki M, Kim KJ, Chung JM, Sun BY, Chung MG. (2012) Genetic diversity in the common terrestrial orchid *Oreorchis patens* and its rare congener *Oreorchis coreana*: Inference of species evolutionary history and implications for conservation. *Journal of Heredity* **103**(5), 692–702.
Chung MY, Nason JD, Chung MG. (2005) Spatial genetic structure in populations of the terrestrial orchid *Orchis cyclochila* (Orchidaceae). *Plant Systematics and Evolution* **254**(3–4), 209–219.
Clements MA, Muir H, Cribb PJ. (1986) A preliminary report on the symbiotic germination of European terrestrial orchids. *Kew Bulletin* **41**(2), 437–445.

Coates F, Lunt I, Tremblay RL. (2006) The effect of fire, rainfall, herbivory on population dynamics of a threatened terrestrial orchid from south-eastern Australia, *Prasophyllum correctum* DL and consequences for management. *Biological Conservation* **129**, 59–69.

Cozzolino S, Nardella AM, Impagliazzo S, Widmer A, Lexer C. (2006) Hybridization and conservation of Mediterranean orchids: Should we protect the orchid hybrids or the orchid hybrid zones? *Biological Conservation* **129**(1), 14–23.

Cozzolino S, Noce ME, Musacchio A, Widmer A. (2003) Variation at a chloroplast minisatellite locus reveals the signature of habitat fragmentation and genetic bottlenecks in the rare orchid *Anacamptis palustris* (Orchidaceae). *American Journal of Botany* **90**(12), 1681–1687.

Cozzolino S, Widmer A. (2005) Orchid diversity: an evolutionary consequence of deception? *Trends in Ecology & Evolution* **20**(9), 487–494.

Cruden RW. (1977) Pollen-ovule ratios: a conservative indicator of breeding systems in flowering plants. *Evolution* **31**(1), 32–46.

Currah RS, Zelmer CD, Hambleton S, Richardson KA. (1997) Fungi from orchid mycorrhizas. In *Orchid Biology: Reviews and Perspectives, VII* (eds. J Arditti and AM Pridgeon) pp. 117–170. Kluwer Academic Publishers: Dordrecht.

Curtis JT. (1939) The relationships of specificity of orchid mycorrhizal fungi to the problem of symbiosis. *American Journal of Botany* **26**, 390–399.

Dafni A, Calder DM. (1987) Pollination by deceit and floral mimesis in *Thelymitra antennifera* (Orchidaceae). *Plant Systematics and Evolution* **158**, 11–22.

Darwin CR. (1862) *On the various contrivances by which British and foreign orchids are fertilised by insects, and on the good effects of intercrossing.* John Murray: London.

Davis BJ, Phillips RD, Wright M, Linde CC, Dixon KW. (2015) Continent-wide distribution in mycorrhizal fungi: implications for the biogeography of specialized orchids. *Annals of Botany* **116**(3), 413–421.

de Boer HJ, Ouarghidi A, Martin G, Abbad A, Kool A. (2014) DNA barcoding reveals limited accuracy of identifications based on folk taxonomy. *PloS One* **9**(1), 9:e84291.

de Capriles CH, Mata S, Middelveen M. (1989) Preservation of fungi in water (Castellani): 20 years. *Mycopathologia* **106**(2), 73–79.

Dearnaley JDW. (2007) Further advances in orchid mycorrhizal research. *Mycorrhiza* **17**(6), 475–486.

Deguilloux M, Pemonge M, Petit R. (2002) Novel perspectives in wood certification and forensics: dry wood as a source of DNA. *Proceedings of the Royal Society of London B: Biological Sciences* **269**(1495), 1039–1046.

Deng H, Zhang GQ, Lin M, Wang Y, Liu ZJ. (2015) Mining from transcriptomes: 315 single-copy orthologous genes concatenated for the phylogenetic analyses of Orchidaceae. *Ecology and evolution* **5**(17), 3800–3807.

DeNiro MJ, Epstein S. (1978) Influence of diet on the distribution of carbon isotopes in animals. *Geochimica et cosmochimica acta* **42**(5), 495–506.

Devey DS, Bateman RM, Fay MF, Hawkins JA. (2008) Friends or relatives? Phylogenetics and species delimitation in the controversial European orchid genus *Ophrys*. *Annals of Botany* **101**(3), 385.

Diez JM. (2007) Hierarchical patterns of symbiotic orchid germination linked to adult proximity and environmental gradients. *Journal of Ecology* **95**(1), 159–170.

Ding G, Zhang D, Ding X, Zhou Q, Zhang W, Li X. (2008) Genetic variation and conservation of the endangered Chinese endemic herb *Dendrobium officinale* based on SRAP analysis. *Plant Systematics and Evolution* **276**(3–4), 149–156.

Dixon KW. (1987) Raising terrestrial orchids from seed. In *Orchid Growing. Proceedings of the 10th Australian Orchid Conference*, Adelaide, Australia.

Dixon KW. (1991) Seeder/clonal concepts in Western Australian orchids. In *Population ecology of terrestrial orchids* (eds. TCE Wells and JH Willems) pp. 111–123. SPB Academic Publishing: The Hague, The Netherlands.

Dixon KW. (2009) Pollination and restoration. *Science* **325**(5940), 571–573.

Dixon KW, Barrett R. (2003) Fire, soils and plant nutrition in West Australian forests: a focus on spatial and temporal variability. In *Fire in south-western Australian ecosystems: Impacts and management* (eds. N Burrows and I Abbott) pp. 205–223. Backhuys Publishers: Leiden, The Netherlands.

Dixon KW, Hopper SD. (1996) Regional accounts—Australia. *In Orchids—status, survey and conservation action plan* (eds. E Hagsater and V Dumont) pp. 109–118. IUCN/SSC Orchid Specialist Group: Gland, Switzerland.

Doyle JJ, Doyle JL. (1987) A rapid DNA isolation procedure from small quantities of fresh leaf tissue. *Phytochemistry Bulletin* **19**, 11–15.

Dressler RL. (2005) How many orchid species? *Selbyana, The Journal of the Marie Selby Botanical Gardens* **26**(1), 155–158.

Duffy KJ, Fay MF, Smith RJ, Stout JC. (2011) Population genetics and conservation of the small white orchid, *Pseudorchis albida,* in Ireland. *Biology and Environment-Proceedings of the Royal Irish Academy* **111B**(2), 73–81.

Engelmann F. (2004) Plant cryopreservation: progress and prospects. *In Vitro Cellular & Developmental Biology-Plant* **40**(5), 427–433.

Ercole E, Adamo M, Rodda M, Gebauer G, Girlanda M, Perotto S. (2015) Temporal variation in mycorrhizal diversity and carbon and nitrogen stable isotope abundance in the wintergreen meadow orchid *Anacamptis morio. New Phytologist* **205**(3), 1308–1319.

Eum SM, Gale S, Yukawa T, Lee NS. (2011) Phylogenetic and conservation status of the endangered terrestrial orchid *Nervilia nipponica* (Orchidaceae) in Korea. *Biochemical Systematics and Ecology* **39**(4–6), 635–642.

Excoffier L, Heckel G. (2006) Computer programs for population genetics data analysis: a survival guide. *Nature Reviews Genetics* **7**(10), 745–758.

Excoffier L, Laval G, Schneider S. (2005) Arlequin (version 3.0): an integrated software package for population genetics data analysis. *Evolutionary Bioinformatics Online* **1**, 47.

Falk DA, Holsinger KE (eds.). (1991) *Genetics and conservation of rare plants.* Oxford University Press: New York, USA.

Farnsworth E, Klionsky S, Brumback W, Havens K. (2006) A set of simple decision matrices for prioritizing collection of rare plant species for *ex situ* conservation. *Biological Conservation* **128**(1), 1–12.

Fay M, Cowan R. (2001) Plastid microsatellites in *Cypripedium calceolus* (Orchidaceae): genetic fingerprints from herbarium specimens. *Lindleyana* **16**(3), 151–156.

Fay MF, Bone R, Cook P, Kahandawala I, Greensmith J, Harris S, Pedersen HÆ, Ingrouille MJ, Lexer C. (2009) Genetic diversity in *Cypripedium calceolus* (Orchidaceae) with a focus on north-western Europe, as revealed by plastid DNA length polymorphisms. *Annals of Botany* **104**(3), 517–525.

Fay MF, Cowan RS, Leitch IJ. (2005) The effects of nuclear DNA content (C-value) on the quality and utility of AFLP fingerprints. *Annals of Botany* **95**(1), 237–246.

Fay MF, Krauss SL. (2003) Orchid conservation genetics in the molecular age. In *Orchid Conservation* (eds. KW Dixon, SP Kell, RL Barrett, and PJ Cribb) pp. 91–112. Natural History Publications: Kota Kinabalu, Sabah, Malaysia.

Fay MF, Taylor I. (2015) 801. *Cypripedium calceolus. Curtis's Botanical Magazine* **32**(1), 24–32.

Felsenstein J. (1989) PHYLIP-phylogeny inference package (version 3.2). *Cladistics* **5**(1), 164–166.

Felsenstein J, Felenstein J. (2004) *Inferring phylogenies.* Sinauer Associates Sunderland: Massachusetts, USA.

Filion M, St-Arnaud M, Jabaji-Hare SH. (2003) Direct quantification of fungal DNA from soil substrate using real-time PCR. *Journal of Microbiological Methods* **53**(1), 67–76.

Flachsland E, Terada G, Scocchi A, Rey H, Mroginski L, Engelmann F. (2006) Cryopreservation of seeds and *in vitro*-cultured protocorms of *Oncidium bifolium* sims. (Orchidaceae) by encapsulation-dehydration. *CryoLetters* **27**(4), 235–242.

Flanagan NS, Peakall R, Clements MA, Otero JT. (2006) Conservation of taxonomically difficult species: the case of the Australian orchid, *Microtis angusii. Conservation Genetics* **7**(6), 847–859.

Fochi V, Chitarra W, et al. (2016) Fungal and plant gene expression in the *Tulasnella calospora–Serapias vomeracea* symbiosis provides clues about nitrogen pathways in orchid mycorrhizas. *New Phytologist*, In press.

Forrest AD, Hollingsworth ML, Hollingsworth PM, Sydes C, Bateman RM. (2004) Population genetic structure in European populations of *Spiranthes romanzoffiana* set in the context of other genetic studies on orchids. *Heredity* **92**(3), 218–227.

Franke S, Ibarra F, Schulz CM, Twele R, Poldy J, Barrow RA, Peakall R, Schiestl FP, Francke W. (2009) The discovery of 2,5-dialkylcyclohexan-1,3-diones as a new class of natural products. *Proceedings of the National Academy of Sciences of the United States of America* **106**(22), 8877–8882.

Frankham R, Ballou JD, Briscoe DA. (2002) *Introduction to conservation genetics.* Cambridge University Press: Cambridge, UK.

Freudenstein JV, Senyo DM. (2008) Relationships and evolution of matK in a group of leafless orchids (*Corallorhiza* and *Corallorhizinae*; Orchidaceae: Epidendroideae). *American Journal of Botany* **95**(4), 498–505.

Galen C. (1999) Why do flowers vary? *Bioscience* **49**(8), 631–640.

Gams W. (2002) *Ex situ* conservation of microbial diversity. In *Microorganisms in Plant Conservation and Biodiversity*. Kluwer: The Netherlands.

Gardes M, Bruns T. (1993) ITS primers with enhanced specificity for basidiomycetes—application to the identification of mycorrhizae and rusts. *Molecular Ecology* **2**(2), 113–118.

Garnett R, Collier P. (2012) It's a myth: Touching orchid plants does not affect their predation. *The Tasmanian Naturalist* (134), 16–23.

Gaskett A, Endler J, Phillips R. (2016) Convergent evolution of sexual deception via chromatic and achromatic contrast rather than colour mimicry. *Evolutionary Ecology*, 1–23.

Gaskett AC. (2011) Orchid pollination by sexual deception: pollinator perspectives. *Biological Reviews* **86**(1), 33–75.

Gebauer G, Dietrich P. (1993) Nitrogen isotope ratios in different compartments of a mixed stand of spruce, larch and beech trees and of understorey vegetation including fungi. *Isotopes in Environmental and Health Studies* **29**(1–2), 35–44.

Gebauer G, Meyer M. (2003) N-15 and C-13 natural abundance of autotrophic and mycoheterotrophic orchids provides insight into nitrogen and carbon gain from fungal association. *New Phytologist* **160**(1), 209–223.

Gebauer G, Preiss K, Gebauer AC. (2016) Partial mycoheterotrophy is more widespread among orchids than previously assumed. *New Phytologist.*

Ghorbani A, Gravendeel B, Naghibi F, de Boer H. (2014) Wild orchid tuber collection in Iran: a wake-up call for conservation. *Biodiversity and Conservation* **23**(11), 2749–2760.

Ghorbani A, Zarre S, Gravendeel B, de Boer HJ. (2014) Illegal wild collection and international trade of CITES-listed terrestrial orchid tubers in Iran. *Traffic Bulletin* **26**(2), 52–58.

Giovannetti M, Mosse B. (1980) An evaluation of techniques for measuring vesicular-arbuscular infection in roots. *New Phytologist* **84**, 489–500.

Girlanda M, Segreto R, Cafasso D, Liebel HT, Rodda M, Ercole E, Cozzolino S, Gebauer G, Perotto S. (2011) Photosynthetic Mediterranean meadow orchids feature partial mycoheterotrophy and specific mycorrhizal associations. *American Journal of Botany* **98**(7), 1148–1163.

Gleixner G, Danier H-J, Werner RA, Schmidt H-L. (1993) Correlations between the 13C content of primary and secondary plant products in different cell compartments and that in decomposing basidiomycetes. *Plant Physiology* **102**(4), 1287–1290.

Goh KM, Haynes R. (1977) Evaluation of potting media for commercial nursery production of container grown plants: III. Effects of media, fertiliser nitrogen, and a nitrification inhibitor on soil nitrification and nitrogen recovery of *Callistephus chinensis* (L.) Nees 'Pink Princess'. *New Zealand Journal of Agricultural Research* **20**(3), 383–393.

Grant V. (1994) Modes and origins of mechanical and ethological isolation in angiosperms. *Proceedings of the National Academy of Sciences of the United States of America* **91**(1), 3–10.

Gregg KB. (1989) Reproductive biology of the orchid *Cleistes divaricata* (L.) Ames var. bifaria Fernald growing in a West Virginia meadow. *Castanea* **54**(2), 57–78.

Gregg KB. (1991) Variation in the behaviour of four populations of the orchid *Cleistes divaricata*, an assessment using transition matrix models. In *Population ecology of terrestrial orchids* (eds. TCE Wells and JH Willems) pp. 139–159. SPB Academic Publishing: The Hague.

Guerrant Jr. EO. (1996) Designing populations: demographic, genetic, and horticultural dimensions. In *Restoring diversity: strategies for reintroduction of endangered plants* (eds. DA Falk, CI Millar, and MO Lwell) pp. 171–208. Island Press: Washington, DC.

Guindon S, Dufayard J-F, Lefort V, Anisimova M, Hordijk W, Gascuel O. (2010) New algorithms and methods to estimate maximum-likelihood phylogenies: assessing the performance of PhyML 3.0. *Systematic Biology* **59**(3), 307–321.

Guo YY, Luo YB, Liu ZJ, Wang XQ. (2012) Evolution and biogeography of the slipper orchids: Eocene vicariance of the conduplicate genera in the old and new world tropics. *Plos One* 7(6).

Gustafsson S. (2000) Patterns of genetic variation in *Gymnadenia conopsea*, the fragrant orchid. *Molecular Ecology* **9**(11), 1863–1872.

Gustafsson S. (2003) *Population genetic analyses in the orchid genus Gymnadenia: a conservation genetic perspective.* Uppsala University, Interfaculty Units, Acta Universitatis Upsaliensis.

Gustafsson S, Lonn M. (2003) Genetic differentiation and habitat preference of flowering-time variants within *Gymnadenia conopsea. Heredity* **91**, 284–292.

Hadley G. (1982) Orchid mycorrhiza. In *Orchid biology: Reviews and perspectives, II. Vol. 2* (ed. J Arditti) pp. 83–181. Cornell University Press: Ithaca, New York, USA.

Hamrick JL, Godt MJW. (1996) Effects of life history traits on genetic diversity in plant species. *Philosophical Transactions of the Royal Society of London B Biological Sciences* **351**(1345), 1291–1298.

Hardy GES, Sivasithamparam K. (1989) Microbial, chemical and physical changes during composting of a eucalyptus (*Eucalyptus calophylla* and *Eucalyptus diversicolor*) bark mix. *Biology and Fertility of Soils* **8**, 260–270.

Hardy GES, Sivasithamparam K. (1991) Suppression of *Pytophtora* root in a composted *Eucalyptus* bark mix. *Australian Journal of Botany* **39**, 153–159.

Harley JL, Smith SE. (1983) *Mycorrhizal symbiosis.* Academic Press: London.

Harvais G, Hadley G. (1967) The relation between host and endophyte in orchid mycorrhiza. *New Phytologist* **66**, 205–215.

Harvais G, Raitsakas A. (1975) On the physiology of a fungus symbiotic with orchids. *Canadian Journal of Botany* **53**, 144–155.

Havens K, Vitt P, Maunder M, Guerrant Jr EO, Dixon K. (2006) *Ex situ* plant conservation and beyond. *Bioscience* **56**(6), 525–531.

Hebert PD, Cywinska A, Ball SL. (2003) Biological identifications through DNA barcodes. *Proceedings of the Royal Society of London B: Biological Sciences* **270**(1512), 313–321.

Hibert F, Taberlet P, Chave J, Scotti-Saintagne C, Sabatier D, Richard-Hansen C. (2013) Unveiling the diet of elusive rainforest herbivores in next generation sequencing era? The tapir as a case study. *PloS One* **8**(4), e60799.

Hicks AJ. (2000) Asymbiotic technique of orchid seed germination. In *The orchid seedbank project* (ed. R Huber) pp. 134. Chandler: Arizona, USA.

Hill WG. (2010) Understanding and using quantitative genetic variation. *Philosophical Transactions of the Royal Society of London B Biological Sciences* **365**(1537), 73–85.

Hobbs RJ, Cramer VA. (2008) Restoration ecology: Interventionist approaches for restoring and maintaining ecosystem function in the face of rapid environmental change. *Annual Review of Environment and Resources* **33**(1), 39–61.

Hogbin PM, Peakall R, Sydes MA. (2000) Achieving practical outcomes from genetic studies of rare Australian plants. *Australian Journal of Botany* **48**, 375–382.

Hoitink HAJ. (1980) Composted bark, a lightweight growth medium with fungicidal properties. *Plant Disease* **64**, 142–147.

Hopper SD. (2000) How well do phylogenetic studies inform the conservation of Australian plants? *Australian Journal of Botany* **48**, 321–328.

Howes BJ, Pither R, Prior KA. (2009) Conservation implications should guide the application of conservation genetics research. *Endangered Species Research* **8**, 193–199.

Hsu RCC, Lee YI. (2012) Seed development of *Cypripedium debile* Rchb. f. in relation to asymbiotic germination. *Hortscience* **47**(10), 1495–1498.

Hu C, Tian H, Li H, Hu A, Xing F, Bhattacharjee A, Hsu T, Kumar P, Chung S. (2016) Phylogenetic analysis of a 'Jewel Orchid' genus *Goodyera* (Orchidaceae) based on DNA sequence data from nuclear and plastid regions. *PloS One* **11**(2), e0150366.

Hufford KM, Mazer SJ. (2003) Plant ecotypes: genetic differentiation in the age of ecological restoration. *Trends in Ecology & Evolution* **18**(3), 147–155.

Husband BC, Campbell LG. (2004) Population responses to novel environments: implications for *ex situ* plant conservation. In *Ex situ plant conservation. Supporting species survival in the wild* (eds. EO Guerrant, K Havens, and M Maunder) pp. 231–266. Island Press: Washington DC, USA.

Huynh TT, McLean CB, Coates F, Lawrie AC. (2004) Effect of developmental stage and peloton morphology on success in isolation of mycorrhizal fungi in *Caladenia formosa* (Orchidaceae). *Australian Journal of Botany* **52**(2), 231–241.

Hynson NA, Madsen TP, Selosse M-A, Adam IK, Ogura-Tsujita Y, Roy M, Gebauer G. (2013) The physiological ecology of mycoheterotrophy. In *Mycoheterotrophy* (ed. V Merckx) pp. 297–342. (Springer: Berlin, Germany.

Hynson NA, Schiebold JM-I, Gebauer G. (2016) Plant family identity distinguishes patterns of carbon and nitrogen stable isotope abundance and nitrogen concentration in mycoheterotrophic plants associated with ectomycorrhizal fungi. *Annals of Botany*. In press.

Hynson NA, Weiss M, Preiss K, Gebauer G, Treseder KK. (2013) Fungal host specificity is not a bottleneck for the germination of *Pyroleae* species (Ericaceae) in a Bavarian forest. *Molecular Ecology* **22**(5), 1473–1481.

Indsto JO, Weston PH, Clements MA, Dyer AG, Batley M, Whelan RJ. (2006) Pollination of *Diuris maculata* (Orchidaceae) by male *Trichocolletes venustus* bees. *Australian Journal of Botany* **54**(7), 669–679.

Indsto JO, Weston PH, Clements MA, Dyer AG, Batley M, Whelan RJ. (2007) Generalised pollination of *Diuris alba* (Orchidaceae) by small bees and wasps. *Australian Journal of Botany* **55**(6), 628–634.

Indsto JO, Weston PH, Clements MA, Whelan RJ. (2005) Corrigendum to: Highly sensitive DNA fingerprinting of orchid pollinaria remnants using AFLP. *Australian Systematic Botany* **18**(5), 473.

IUCN. (1987) IUCN position statement on the translocation of living organisms: Introductions, re-introductions, and re-stocking. Prepared by the Species Survival Commission in collaboration with the Commission on Ecology and the Commission on Environmental Policy, Law and Administration. 22nd Meeting of the IUCN Council, Gland, Switzerland.

IUCN. (1999) IUCN guidelines for the prevention of biodiversity loss due to biological invasion. *Species* **31–32**, 28–42.

Izawa T, Kawahara T, Takahashi H. (2007) Genetic diversity of an endangered plant, Cypripedium macranthos var. rebunense (Orchidaceae): background genetic research for future conservation. *Conservation Genetics* **8**(6), 1369–1376.

Jacquemyn H, Brys R, Cammue BP, Honnay O, Lievens B. (2011) Mycorrhizal associations and reproductive isolation in three closely related *Orchis* species. *Annals of Botany* **107**(3), 347–356.

Jacquemyn H, Brys R, Merckx VS, Waud M, Lievens B, Wiegand T. (2014) Coexisting orchid species have distinct mycorrhizal communities and display strong spatial segregation. *New Phytologist* **202**(2), 616–627.

Jacquemyn H, Deja A, De Hert K, Bailarote BC, Lievens B. (2012) Variation in mycorrhizal associations with Tulasnelloid fungi among populations of five *Dactylorhiza* species. *Plos One* 7(8).

Jacquemyn H, Merckx V, Brys R, Tyteca D, Cammue BPA, Honnay O, Lievens B. (2011) Analysis of network architecture reveals phylogenetic constraints on mycorrhizal specificity in the genus *Orchis* (Orchidaceae). *New Phytologist* **192**(2), 518–528.

Jersakova J, Johnson SD. (2006) Lack of floral nectar reduces self-pollination in a fly-pollinated orchid. *Oecologia* **147**(1), 60–68.

Jersakova J, Johnson SD, Kindlmann P. (2006) Mechanisms and evolution of deceptive pollination in orchids. *Biological Reviews of the Cambridge Philosophical Society* **81**(2), 219–235.

Jesche JM, Strayer DL. (2008) Are threat status and invasion success two sides of the same coin? *Ecography* **31**, 124–130.

Jobes DV, Hurley DL, Thien LB. (1995) Plant DNA isolation: a method to efficiently remove polyphenolics, polysaccharides, and RNA. *Taxon* **44**, 379–386.

Johansson VA, Mikusinska A, Ekblad A, Eriksson O. (2015) Partial mycoheterotrophy in *Pyroleae*: nitrogen and carbon stable isotope signatures during development from seedling to adult. *Oecologia* **177**(1), 203–211.

Johnson SD. (2000) Batesian mimicry in the non-rewarding orchid *Disa pulchra*, and its consequences for pollinator behaviour. *Biological Journal of the Linnean Society* **71**(1), 119–132.

Johnson SD. (2005) Specialized pollination by spider-hunting wasps in the African orchid *Disa sankeyi*. *Plant Systematics and Evolution* **251**(2–4), 153–160.

Johnson SD, Peter CI, Agren J. (2004) The effects of nectar addition on pollen removal and geitonogamy in the non-rewarding orchid *Anacamptis morio*. *Proceedings of the Royal Society of London B Biological Sciences* **271**(1541), 803–809.

Johnson SD, Peter CI, Nilsson LA, Agren J. (2003) Pollination success in a deceptive orchid is enhanced by co-occurring rewarding magnet plants. *Ecology* **84**(11), 2919–2927.

Jones DL. (1991) New species of Australian Orchidaceae. In *Australian Orchid Research. Vol. 2*. pp. 1–208.

Jones KN. (2001) Pollinator-mediated assortative mating: causes and consequences. In *Cognitive ecology of pollination* (eds. L Chittka and JD Thomson) pp. 259–273. Cambridge University Press: Cambridge.

Kasparek M, Grimm U. (1999) European trade in Turkish salep with special reference to Germany. *Economic Botany* **53**(4), 396–406.

Kennedy PG, Bergemann SE, Hortal S, Bruns TD. (2007) Determining the outcome of field-based competition between two *Rhizopogon* species using real-time PCR. *Molecular Ecology* **16**(4), 881–890.

Kindlmann P, Balounova Z. (1999) Energy partitioning in terrestrial orchids—a model for assessing their performance. *Ecological Modelling* **119**, 167–176.

Kirkpatrick M, Barton NH. (1997) Evolution of a species' range. *The American Naturalist* **150**(1), 1–23.

Knudson L. (1927) Symbiosis and asymbiosis relative to orchids. *New Phytologist* **26**, 328–336.

Knudson L. (1954) Storage and viability of orchid seed. *American Orchid Society Bulletin* **22**, 260.

Kool A, de Boer HJ, KrügerÅ, Rydberg A, Abbad A, Björk L, Martin G. (2012) Molecular identification of commercialized medicinal plants in southern Morocco. *PloS One* 7(6), e39459.

Kormanik PP, McGraw AC. (1982) Quantification of vesicular-arbuscular mycorrhizae in plant roots. In *Methods and principles of mycorrhizal research* (ed. C Schenck) pp. 37–45. American Phytopathological Society: St. Paul, Minnesota, USA.

Koske RE, Testier B. (1983) A convenient permanent slide mounting medium. *Mycological Society of America Newsletter* **34**, 59.

Kramer AT, Havens K. (2009) Plant conservation genetics in a changing world. *Trends in Plant Science* **14**(11), 599–607.

Kreziou A, de Boer H, Gravendeel B. (2015) Harvesting of salep orchids in north-western Greece continues to threaten natural populations. *Oryx*, 1–4.

Kuiter RH. (2013) *Orchid pollinators of Victoria Second Edition.* Aquatic Photographics: Seaford, Victoria.

Labate JA. (2000) Software for population genetic analyses of molecular marker data. *Crop Science* **40**(6), 1521.

Lakon G. (1949) The topographical tetrazolium method for determining the germination capacity of the seed. *Plant Physiology* **24**, 389–394.

Landeweert R, Leeflang P, Smit E, Kuyper T. (2005) Diversity of an ectomycorrhizal fungal community studied by a root tip and total soil DNA approach. *Mycorrhiza* **15**(1), 1–6.

Lauzer D, St-Arnaud M, Barabe D. (1994) Tetrazolium staining and *in vitro* germination of mature seeds of *Cypripedium acaule* (Orchidaceae). *Lindleyana* **9**(3), 197–204.

Leake JR. (1994) Tansley review No. 69. The biology of myco-heterotrophic ('saprophytic') plants. *New Phytologist* **127**(2), 171–216.

Lee Y-I, Chung M-C, Yeung EC, Lee N. (2015) Dynamic distribution and the role of abscisic acid during seed development of a lady's slipper orchid, *Cypripedium formosanum. Annals of Botany* (116), 403–411.

Lee Y-I, Lee N, Yeung EC, Chung M-C. (2005) Embryo development of *Cypripedium formosanum* in relation to seed germination *in vitro. Journal of the American Society for Horticultural Science* **130**(5), 747–753.

Lee Y-I, Yang C-K, Gebauer G. (2015) The importance of associations with saprotrophic non-Rhizoctonia fungi among fully mycoheterotrophic orchids is currently under-estimated: novel evidence from sub-tropical Asia. *Annals of Botany*(116), 423–435.

Leles B, Chaves AV, Russo P, Batista JA, Lovato MB. (2015) Genetic structure is associated with phenotypic divergence in floral traits and reproductive investment in a high-altitude orchid from the Iron Quadrangle, southeastern Brazil. *PloS One* **10**(3), e0120645.

Li A, Ge S. (2006) Genetic variation and conservation of Changnienia amoena, an endangered orchid endemic to China. *Plant Systematics and Evolution* **258**(3–4), 251–260.

Li X, Ding X, Chu B, Zhou Q, Ding G, Gu S. (2008) Genetic diversity analysis and conservation of the endangered Chinese endemic herb Dendrobium officinale Kimura et Migo (Orchidaceae) based on AFLP. *Genetica* **133**(2), 159–166.

Liebel HT, Bidartondo MI, Gebauer G. (2015) Are carbon and nitrogen exchange between fungi and the orchid *Goodyera repens* affected by irradiance? *Annals of Botany* **115**(2), 251–261.

Light M, MacConaill M. (2007) Effects of trampling on a terrestrial orchid environment. *Lankesteriana* 7(1–2), 294–298.

Light MHS, MacConaill M. (1990) Characterization of the optimal capsule development stage for embryo culture of *Cypripedium calceolus* var. *pubescens*. In *North American Native Terrestrial Orchid Propagation and Production Conference Proceedings*. pp. 92. Chadds Ford, Pennsylvania.

Light MHS, MacConaill M. (1998) Factors affecting germinable seed yield in *Cypripedium calceolus* var. *pubescens* (Willd.) Correll and *Epipactis helleborine* (L) Crantz (Orchidaceae). *Botanical Journal of the Linnean Society* **126**(1–2), 3–26.

Light MHS, MacConaill M. (2008) Watch your step. *American Orchid Society, Orchids Magazine* **02**, 128–133.

Linder HP. (1995) Setting conservation priorities: The importance of endemism and phylogeny in the southern African orchid genus *Herschelia*. *Conservation Biology* **9**(3), 585–595.

Linington SH, Pritchard HW. (2001) Gene banks. In *Encyclopedia of Biodiversity*, Volume 3 (ed. SA Levin) pp. 165–181. Academic Press: San Diego, USA.

Liu Q, Chen J, Corlett RT, Fan X, Yu D, Yang H, Gao J. (2015) Orchid conservation in the biodiversity hotspot of southwestern China. *Conservation Biology* **29**(6), 1563–1572.

Lombardo VT, Hopkins SE, Selosse MA, Cozzolino S, Taylor DL. (2008) Isolation and characterization of new polymorphic microsatellite loci in the mixotrophic orchid *Limodorum abortivum* L. Swartz (Orchidaceae). *Molecular Ecology Resources* **8**(5), 1117–1120.

Luca A, Bellusci F, Pellegrino G. (2014) Interactions with mycorrhizal fungi in two closely related hybridizing orchid species. *Acta Botanica Croatica* **73**(1), 209–220.

Luo J, Hou B-W, Niu Z-T, Liu W, Xue Q-Y, Ding X-Y. (2014) Comparative chloroplast genomes of photosynthetic orchids: insights into evolution of the Orchidaceae and development of molecular markers for phylogenetic applications. *PloS One* **9**(6), e99016.

Ma X, Shi J, Bänziger H, Sun Y, Guo Y, Liu Z, Johnson SD, Luo Y. (2015) The functional significance of complex floral colour pattern in a food-deceptive orchid. *Functional Ecology*.

Machon N, Bardin P, Mazer SJ, Moret J, Godelle B, Austerlitz F. (2003) Relationship between genetic structure and seed and pollen dispersal in the endangered orchid *Spiranthes spiralis*. *New Phytologist* **157**(3), 677–687.

Mackay TFC, Stone EA, Ayroles JF. (2009) The genetics of quantitative traits: challenges and prospects. *Nature Reviews Genetics* **10**(8), 565–577.

Malmgren S. (1993) Large-scale asymbiotic propagation by seed of *Anacamptis, Ophrys, Orchis* and other European orchids with entire tubers. *Svensk Botanisk Tidskrift* **87**(4), 221–234.

Mant J, Peakall R, Schiestl FP. (2005) Does selection on floral odor promote differentiation among populations and species of the sexually deceptive orchid genus *Ophrys*? *Evolution* **59**(7), 1449–1463.

Martos F, Munoz F, Pailler T, Kottke I, Gonneau C, Selosse MA. (2012) The role of epiphytism in architecture and evolutionary constraint within mycorrhizal networks of tropical orchids. *Molecular Ecology* **21**(20), 5098–5109.

Masuhara G, Katsuya K. (1994) *In situ* and *in vitro* specificity between Rhizoctonia Spp and *Spiranthes sinensis* (Persoon) *ames* Var *amoena* (M Bieberstein) *hara* (Orchidaceae). *New Phytologist* **127**(4), 711–718.

Mattila E. (2000) The effects of water stress and pollen availability on reproductive success of *Platanthera biflora* (Orchidaceae). *Det Norske Videnskaps-Akademi. I. Matematisk Naturvidenskapelige Klasse, Skrifter, Ny Serie* **39,** 83–90.

Mattila E, Kuitunen MT. (2000) Nutrient versus pollination limitation in *Platanthera bifolia* and *Dactylorhiza incarnata* (Orchidaceae). *Oikos* **89**(2), 360–366.

Maunder M. (1992) Plant reintroduction: an overview. *Biodiversity and Conservation* **1**, 51–61.

Maunder M, Higgens S, Culham A. (2001) The effectiveness of botanic garden collections in supporting plant conservation: a European case study. *Biodiversity and Conservation* **10**(3), 383–401.

McCormick MK, Taylor DL, Juhaszova K, Burnett RK, Whigham DF, O'Neill JP. (2012) Limitations on orchid recruitment: not a simple picture. *Molecular Ecology* **21**(6), 1511–1523.

McCormick MK, Whigham DF, O'Neill J. (2004) Mycorrhizal diversity in photosynthetic terrestrial orchids. *New Phytologist* **163**(2), 425–438.

McGonigle TP, Miller MH, Evans DG, Fairchild GL, Swan JA. (1990) A new method which gives an objective measure of colonisation of roots by vesicular-arbuscular mycorrhizal fungi. *New Phytologist* **115**, 495–501.

Mckendrick SL. (1995) The effects of herbivory and vegetation on laboratory-raised *Dactylorhiza praetermissa* (Orchidaceae) planted into grassland in southern England. *Biological Conservation* **73**(3), 215–220.

McKendrick SL, Leake JR, Taylor DL, Read DJ. (2002) Symbiotic germination and development of the myco-heterotrophic orchid *Neottia nidus-avis* in nature and its requirement for locally distributed *Sebacina* spp. *New Phytologist* **154**(1), 233–247.

Mehrhoff LA. (1983) Pollination in the genus *Isotria* (Orchidaceae). *American Journal of Botany*, 1444–1453.

Menz MH, Phillips RD, Anthony JM, Bohman B, Dixon KW, Peakall R. (2015) Ecological and genetic evidence for cryptic ecotypes in a rare sexually deceptive orchid, *Drakaea elastica. Botanical Journal of the Linnean Society* **177**(1), 124–140.

Menz MH, Phillips RD, Winfree R, Kremen C, Aizen MA, Johnson SD, Dixon KW. (2011) Reconnecting plants and pollinators: challenges in the restoration of pollination mutualisms. *Trends in plant science* **16**(1), 4–12.

Merritt DJ, Hay FR, Swarts ND, Sommerville KD, Dixon KW. (2014) Ex Situ Conservation and Cryopreservation of Orchid Germplasm. *International Journal of Plant Sciences* **175**(1), 46–58.

Micheneau C, Carlsward BS, Fay MF, Bytebier B, Pailler T, Chase MW. (2008) Phylogenetics and biogeography of Mascarene angraecoid orchids (Vandeae, Orchidaceae). *Molecular phylogenetics and evolution* **46**(3), 908–922.

Micheneau C, Johnson SD, Fay MF. (2009) Orchid pollination: from Darwin to the present day. *Botanical Journal of the Linnean Society* **161**(1), 1–19.

Miller MP. (1997) Tools for population genetic analyses (TFPGA) 1.3: A Windows program for the analysis of allozyme and molecular population genetic data. Department of Biological Sciences, Box 5640, Northern Arizona University, Flagstaff, AZ 86011-5640.

Minasiewicz J, Znaniecka JM. (2014) Characterization of 15 novel microsatellite loci for *Cypripedium calceolus* (Orchidaceae) using MiSeq sequencing. *Conservation Genetics Resources* **6**(3), 527–529.

Miyoshi K, Mii M. (1998) Stimulatory effects of sodium and calcium hypochlorite, pre-chilling and cytokinins on the germination of *Cypripedium macranthos* seed in vitro. *Physiologia Plantarum* **102**(4), 481–486.
Molvray M, Kores PJ, Chase MW. (2000) Polyphyly of mycoheterotropic orchids and functional influences on floral and molecular characteristics. In *Monocots: systematics and evolution* (eds. KL Wilson and DA Mossison) pp. 441–448. CSIRO: Melbourne, Australia.
Montalvo AM, Ackerman JD. (1987) Limitations to fruit production in *Ionopsis utricularioides* (Orchidaceae). *Biotropica* **19**(1), 24–31.
Moore RT. (1987) The genera of *Rhizoctonia*-like fungi: *Aschorhizoctonia, Ceratorhiza* gen. nov., *Epulorhiza* gen nov., *Moniliopsis* and *Rhizoctonia*. *Mycotaxon* **29**, 91–99.
Morrison DA. (2005) Networks in phylogenetic analysis: new tools for population biology. *International Journal for Parasitology* **35**(5), 567–582.
Muir HJ. (1989) Germination and mycorrhizal fungus compatibility in European orchids. In *Modern methods in orchid conservation: The role of physiology, ecology and management* (ed. HW Pritchard) pp. 39–56. Cambridge University Press: Cambridge, UK.
Münzbergová Z, Herben T. (2005) Seed, dispersal, microsite, habitat and recruitment limitation: identification of terms and concepts in studies of limitations. *Oecologia* **145**(1), 1–8.
Myers N, Mittermeier RA, Mittermeier CG, da Fonseca GAB, Kent J. (2000) Biodiversity hotspots for conservation priorities. *Nature* **403**, 853–858.
Nash N, Barrett RL, Oakeley HF, Chalmers I, Richards H. (2003) Role of orchid societies and growers in conservation. In *Orchid Conservation* (eds. KW Dixon, SP Kell, RL Barrett, and PJ Cribb) pp. 313–328. Natural History Publications Kota Kinabalu, Sabah, Malaysia.
Nature IUfCo, Resources N, Commission ISS. (2001), *IUCN Red List categories and criteria*. IUCN.
Neigel JE. (1997) A comparison of alternative strategies for estimating gene flow from genetic markers. *Annual Review of Ecology and Systematics* **28**, 105–128.
Neiland MRM, Wilcock CC. (1998) Fruit set, nectar reward, and rarity in the Orchidaceae. *American Journal of Botany* **85**(12), 1657–1671.
Neiland MRM, Wilcock CC. (1999) The presence of heterospecific pollen on stigmas of nectariferous and nectarless orchids and its consequences for their reproductive success. *Protoplasma* **208**(1–4), 65–75.
Newman EI. (1966) A method of estimating the total length of root in a sample. *Journal of Applied Ecology* **3**, 139–145.
Newmaster SG, Grguric M, Shanmughanandhan D, Ramalingam S, Ragupathy S. (2013) DNA barcoding detects contamination and substitution in North American herbal products. *BMC Medicine* **11**(1), 1.
Nielsen R, Slatkin M. (2000) Likelihood analysis of ongoing gene flow and historical association. *Evolution* **54**(1), 44–50.
Nilsson L. (1992) Animal pollinators adjust plant gender in relation to floral display: evidence from *Orchis morio* (Orchidaceae). *Evolutionary Trends in Plants* **6**(1), 33–40.
Nilsson LA. (1981) The pollination of *Listera ovata* (Orchidaceae). *Nordic Journal of Botany* **1**, 461–480.

Nomura N, Ogura-Tsujita Y, Gale SW, Maeda A, Umata H, Hosaka K, Yukawa T. (2013) The rare terrestrial orchid *Nervilia nipponica* consistently associates with a single group of novel mycobionts. *Journal of Plant Research* **126**(5), 613–623.

Ogura-Tsujita Y, Gebauer G, Hashimoto T, Umata H, Yukawa T. (2009) Evidence for novel and specialized mycorrhizal parasitism: the orchid *Gastrodia confusa* gains carbon from saprotrophic *Mycena. Proceedings of the Royal Society of London B Biological Sciences* **276**(1657), 761–767.

Oliveira SF, Bocayuva MF, Veloso TGR, Bazzolli DMS, da Silva CC, Pereira OL, Kasuya MCM. (2014) Endophytic and mycorrhizal fungi associated with roots of endangered native orchids from the Atlantic Forest, Brazil. *Mycorrhiza* **24**(1), 55–64.

Otero JT, Ackerman J, Bayman P. (2004) Differences in mycorrhizal preferences between two tropical orchids. *Molecular Ecology* **13**(8), 2393–2404.

Otero JT, Flanagan NS, Herre EA, Ackerman JD, Bayman P. (2007) Widespread mycorrhizal specificity correlates to mycorrhizal function in the neotropical, epiphytic orchid *Ionopsis utricularioides* (Orchidaceae). *American Journal of Botany* **94**(12), 1944.

Paul J, Budd C, Freeland JR. (2013) Conservation genetics of an endangered orchid in eastern Canada. *Conservation Genetics* **14**(1), 195–204.

Paulus HF, Gack C. (1990) Pollinators as prepollinating isolation factors: evolution and speciation in *Ophrys* (Orchidaceae). *Israel Journal of Botany* **39**, 43–79.

Peakall R. (1990) Responses of male *Zaspilothynnus trilobatus* turner wasps to females and the sexually deceptive orchid it pollinates. *Functional Ecology* **4**(2), 159–167.

Peakall R. (2007) Speciation in the Orchidaceae: confronting the challenges. *Molecular Ecology* **16**, 2834–2837.

Peakall R, Beattie AJ. (1996) Ecological and genetic consequences of pollination by sexual deception in the orchid *Caladenia tentactulata. Evolution* **50**(6), 2207–2220.

Peakall R, Handel SN. (1993) Pollinators discriminate among floral heights of a sexually deceptive orchid: Implications for selection. *Evolution* 47(6), 1681–1687.

Peakall R, Smouse PE. (2006) GENALEX 6: genetic analysis in Excel. Population genetic software for teaching and research. *Molecular Ecology Notes* **6**(1), 288–295.

Peakall R, Sydes MA. (1996) Defining priorities for achieving practical outcomes from the genetic studies of rare plants. In *Meeting on back from the brink—refining the threatened species recovery process*, 1996, Sydney, Australia (eds. S Stephens and S Maxwell), pp. 119–129.

Pedersen HÆ, Rasmussen HN, Kahandawala IM, Fay MF. (2012) Genetic diversity, compatibility patterns and seed quality in isolated populations of *Cypripedium calceolus* (Orchidaceae). *Conservation Genetics* **13**(1), 89–98.

Pedersen HÆ, Watthana S, Roy M, Suddee S, Selosse MA. (2009) Cephalanthera exigua rediscovered: new insights in the taxonomy, habitat requirements and breeding system of a rare mycoheterotrophic orchid. *Nordic Journal of Botany* **27**(6), 460–468.

Pellegrino G, Luca A, Bellusci F. (2016) Relationships between orchid and fungal biodiversity: Mycorrhizal preferences in Mediterranean orchids. *Plant Biosystems-An International Journal Dealing with all Aspects of Plant Biology* **150**(2), 180–189.

Pellegrino G, Noce ME, Bellusci F, Musacchio A. (2006) Reproductive biology and conservation genetics of *Serapias vomeracea* (Ochidaceae). *Folia Geobotanica* **41**(1), 21–32.

Pellegrino G, Palermo AM, Noce ME, Bellusci F, Musacchio A. (2007) Genetic population structure in the Mediterranean *Serapias vomeracea*, a nonrewarding orchid group.

Interplay of pollination strategy and stochastic forces? *Plant Systematics and Evolution* **263**(3/4), 145–157.

Perkins AJ, Masuhara G, McGee PA. (1995) Specificity of the associations between *Microtis parviflora* (Orchidaceae) and its mycorrhizal fungi. *Australian Journal of Botany* **43**(1), 85–91.

Perkins AJ, McGee PA. (1995) Distribution of the orchid mycorrhizal fungus, *Rhizoctonia solani*, in relation to its host, *Pterostylis acuminata*, in the field. *Australian Journal of Botany* **43**(6), 565–575.

Peterson RL, Uetake Y, Zelmer C. (1998) Fungal symbioses with orchid protocorms. *Symbiosis* **25**(1–3), 29–55.

Pfeifer M, Heinrich W, Jetschke G. (2006) Climate, size and flowering history determine flowering pattern of an orchid. *Botanical Journal of the Linnean Society* **151**(4), 511–526.

Pfeifer M, Schatz B, Pico FX, Passalacqua NG, Fay MF, Carey PD, Jeltsch F. (2009) Phylogeography and genetic structure of the orchid *Himantoglossum hircinum* (L.) Spreng. across its European central-marginal gradient. *Journal of Biogeography* **36**(12), 2353–2365.

Phillips JM, Hayman DS. (1970) Improved procedures for clearing roots and staining parasitic and vesicular-arbuscular mycorrhizal fungi for rapid assessment of infection. *Transactions of the British Mycological Society* **55**(1), 158–160.

Phillips RD, Barrett MD, Dalziell EL, Dixon KW, Swarts ND. (2016) Geographical range and host breadth of Sebacina orchid mycorrhizal fungi associating with *Caladenia* in south-western Australia. *Botanical Journal of the Linnean Society* **182**(1), 140–151.

Phillips RD, Barrett MD, Dixon KW, Hopper SD. (2011) Do mycorrhizal symbioses cause rarity in orchids? *Journal of Ecology* **99**(3), 858–869.

Phillips RD, Bohman B, Anthony J, Krauss S, Dixon K, Peakall R. (2015) Mismatch in the distribution of floral ecotypes and pollinators: insights into the evolution of sexually deceptive orchids. *Journal of Evolutionary Biology* **28**(3), 601–612.

Phillips RD, Brown AP, Dixon KW, Hopper SD. (2011) Orchid biogeography and factors associated with rarity in a biodiversity hotspot, the Southwest Australian Floristic Region. *Journal of Biogeography* **38**(3), 487–501.

Phillips RD, Dixon KW, Peakall R. (2012) Low population genetic differentiation in the Orchidaceae: implications for the diversification of the family. *Molecular Ecology* **21**, 5208–5220.

Phillips RD, Peakall R, Hutchinson MF, Linde CC, Xu TB, Dixon KW, Hopper SD. (2014) Specialized ecological interactions and plant species rarity: The role of pollinators and mycorrhizal fungi across multiple spatial scales. *Biological Conservation* **169**, 285–295.

Pillon Y, Fay MF, Shipunov AB, Chase MW. (2006) Species diversity versus phylogenetic diversity: A practical study in the taxonomically difficult genus *Dactylorhiza* (Orchidaceae). *Biological Conservation* **129**(1), 4–13.

Pillon Y, Qamaruz-Zaman F, Fay MF, Hendoux F, Piquot Y. (2007) Genetic diversity and ecological differentiation in the endangered fen orchid (*Liparis loeselii*). *Conservation Genetics* **8**(1), 177–184.

Pinheiro F, de Barros F, Palma-Silva C, Meyer D, Fay MF, Suzuki RM, Lexer C, Cozzolino S. (2010) Hybridization and introgression across different ploidy levels in the Neotropical orchids *Epidendrum fulgens* and *E. puniceoluteum* (Orchidaceae). *Molecular Ecology* **19**(18), 3981–3994.

Poldy J, Peakall R, Barrow RA. (2005) Investigations into the wasp (*Neozeleboria cryptoides*)—orchid (*Chiloglottis trapeziformis*) semiochemical relationship. In *The International Chemical Congress of Pacific Basin Societies*, Hawaii.

Porter IJ, Merriman PR. (1985) Evaluation of soil solarization for control of root disease of row crops in Victoria. *Plant Pathology* **34**, 108–118.

Preiss K, Gebauer G. (2008) A methodological approach to improve estimates of nutrient gains by partially myco-heterotrophic plants. *Isotopes in Environmental Health Studies* **44**, 393–401.

Pridgeon AM. (2003) Modern species concepts and practical considerations for conservation of Orchidaceae. In *Orchid Conservation* (eds. KW Dixon, SP Kell, RL Barrett, and PJ Cribb) pp. 43–54. Natural History Publications: Kota Kinabalu, Sabah, Malaysia.

Pritchard A. Current methods and case studies from Eastern Australia on the translocation/transplantation of threatened orchid taxa (Talk). In *Orchid Conservation Forum II*, 2003, Royal Botanic Gardens, Melbourne (ed. R Cross).

Pritchard HW. (1984) Liquid nitrogen preservation of terrestrial and epiphytic orchid seed. *Cryo Letters* **5**, 295–300.

Pritchard HW. (1985) Determination of orchid seed viability using fluorescein diacetate. *Plant, Cell and Environment* **8**, 727–730.

Pritchard HW, Poynter ALC, Seaton PT. (1999) Interspecific variation in orchid seed longevity in relation to ultra-dry storage and cryopreservation. *Lindleyana* **14**(2), 92–101.

Pritchard JK, Stephens M, Donnelly P. (2000) Inference of population structure using multilocus genotype data. *Genetics* **155**(2), 945–959.

Prosser JI, Tough AJ. (1991) Growth mechanisms and growth kinetics of filamentous microorganisms. *Critical Reviews in Biotechnology* **10**(4), 253–274.

Qamaruz-Zaman F, Fay MF, Parker JS, Chase MW. (1998) Molecular techniques employed in the assessment of genetic diversity: a review focusing on orchid conservation. *Lindleyana* **13**(4), 259–283.

Qian X, Li Q-J, Liu F, Gong M-J, Wang C-X, Tian M. (2014) Conservation genetics of an endangered Lady's Slipper Orchid: *Cypripedium japonicum* in China. *International Journal of Molecular Sciences* **15**(7), 11578–11596.

Qian X, Wang C-x, Tian M. (2013) Genetic diversity and population differentiation of *Calanthe tsoongiana*, a rare and endemic orchid in China. *International Journal of Molecular Sciences* **14**(10), 20399–20413.

Ramsay MM, Dixon KW. (2003) Propagation science, recovery and translocation of terrestrial orchids. In *Orchid Conservation* (eds. KW Dixon, S Kell, RL Barrett, and PJ Cribb) pp. 259–288. Natural History Publications: Kota Kinabalu, Sabah, Malaysia.

Ramsay MM, Stewart J. (1998) Re-establishment of the lady's slipper orchid (*Cypripedium calceolus* L.) in Britain. *Botanical Journal of the Linnean Society* **126**(1–2), 173–181.

Ramsay RR, Sivasithamparam K, Dixon KW. (1986) Patterns of infection and endophytes associated with Western Australian orchids. *Lindleyana* **1**(3), 203–214.

Ramsay RR, Sivasithamparam K, Dixon KW. (1987) Anastomosis groups among *Rhizoctonia*-like endophytic fungi in southwestern Australia *Pterostylis species* (Orchidaceae). *Lindleyana* **2**(3), 161–166.

Rasmussen FN. (2000) Ins and outs of orchid phylogeny. In *Monocots: Systematics and evolution. Vol. Kingsley* (Book) (eds. KL Wilson and DA Morrison) pp. 430–435. CSIRO: Melbourne, Australia.

Rasmussen H, Johansen B, Andersen TF. (1989) Density-dependent interactions between seedlings of *Dactylorhiza majalis* (Orchidaceae) in symbiotic *in vitro* culture. *Physiologia Plantarum* **77**(3), 473–478.

Rasmussen H, Rasmussen FN. (1991) Climatic and seasonal regulation of seed plant establishment in *Dactylorhiza majalis* inferred from symbiotic experiments *in vitro*. *Lindleyana* **6**(4), 221–227.

Rasmussen HN. (1992) Seed dormancy patterns in *Epipactis palustris* (Orchidaceae), requirements for germination and establishment of mycorrhiza. *Physiologia Plantarum* **86**(1), 161–167.

Rasmussen HN. (1995) *Terrestrial orchids, from seed to mycotrophic plant*. Cambridge University Press: Cambridge, UK.

Rasmussen HN. (2002) Recent developments in the study of orchid mycorrhiza. *Plant and Soil* **244**(1–2), 149–163.

Rasmussen HN, Kahandawala IM, Fay MF. (2012) Genetic diversity, compatibility patterns and seed quality in isolated populations of *Cypripedium calceolus* (Orchidaceae). *Conservation Genetics* **13**(1), 89–98.

Rasmussen HN, Pedersen HÆ. (2011) *Cypripedium calceolus* germination *in situ*: seed longevity, and dormancy breakage by long incubation and cold winters. *European Journal of Environmental Sciences* **1**(2), 69–70.

Rasmussen HN, Rasmussen FN. (2009) Orchid mycorrhiza: implications of a mycophagous life style. *OIKOS* **118**(3), 334–345.

Rasmussen HN, Whigham DF. (1993) Seed ecology of dust seeds *in situ*—a new study technique and its application in terrestrial orchids. *American Journal of Botany* **80**(12), 1374–1378.

Rasmussen HN, Whigham DF. (2002) Phenology of roots and mycorrhiza in orchid species differing in phototrophic strategy. *New Phytologist* **154**(3), 797–807.

Raymond M, Rousset F. (1995) An exact test for population differentiation. *Evolution* **49**(6), 1280–1283.

Raymond M, Rousset F. (1995) GENEPOP (version 1.3): population genetics software for exact tests and ecumenicism. *Journal of Heredity* **86**, 248–249.

Reeslev M, Kjoller A. (1995) Comparison of biomass dry weights and radial growth rates of fungal colonies on media solidified with different gelling compounds. *Applied and Environmental Microbiology* **61**(12), 4236–4239.

Reiter N, Whitfield J, Pollard G, Bedggood W, Argall M, Dixon K, Davis B, Swarts N. (2016) Orchid re-introductions: an evaluation of success and ecological considerations using key comparative studies from Australia. *Plant Ecology* **217**(1), 81–95.

Renner SS, Waser NM, Ollerton J. (2006) Rewardless flowers in the angiosperms and the role of insect cognition in their evolution. In *Plant-pollinator interactions: from specialization to generalization* (eds. NM Waser and J Ollerton) pp. 123–144. University of Chicago Press: Chicago, USA.

Richardson RT, Lin C-H, Sponsler DB, Quijia JO, Goodell K, Johnson RM. (2015) Application of ITS2 metabarcoding to determine the provenance of pollen collected by honey bees in an agroecosystem. *Applications in Plant Sciences* **3**(1), 1400066.

Rieseberg LH, Willis JH. (2007) Plant speciation. *Science* **317**(5840), 910.

Roberts DL. (2003) Pollination biology: the role of sexual reproduction in orchid conservation. In *Orchid conservation* (eds. KW Dixon, SP Kell, RL Barrett, and PJ Cribb) pp. 113–136. Natural History Publications: Kota Kinabalu, Sabah, Malaysia.

Roberts P. (1999) *Rhizoctonia-forming fungi*. Whitstable Litho Printers Ltd: Whitstable, Kent, Great Britain.

Roche SA, Carter RJ, Peakall R, Smith LM, Whitehead MR, Linde CC. (2010) A narrow group of monophyletic *Tulasnella* (Tulasnellaceae) symbiont lineages are associated with multiple species of *Chiloglottis* (Orchidaceae): Implications for orchid diversity. *American Journal of Botany* **97**(8), 1313–1327.

Rodrigues KF, Kumar SV. (2009) Isolation and characterization of 24 microsatellite loci in *Paphiopedilum rothschildianum*, an endangered slipper orchid. *Conservation Genetics* **10**(1), 127–130.

Ronquist F, Huelsenbeck JP. (2003) MrBayes 3: Bayesian phylogenetic inference under mixed models. *Bioinformatics* (Oxford) **19**(12), 1572–1574.

Ross AA, Travers SE. (2016) The genetic consequences of rarity in the western prairie fringed orchid (*Platanthera praeclara*). *Conservation Genetics* **17**(1), 69–76.

Rossetto M, Weaver PK, Dixon KW. (1995) Use of RAPD analysis in devising conservation strategies for the rare and endangered *Grevillea scapigera* (Proteaceae). *Molecular Ecology* **4**, 321–329.

Roy M, Gonneau C, Rocheteau A, Berveiller D, Thomas JC, Damesin C, Selosse MA. (2013) Why do mixotrophic plants stay green? A comparison between green and achlorophyllous orchid individuals in situ. *Ecological Monographs* **83**(1), 95–117.

Ruibal MP, Peakall R, Foret S, Linde CC. (2014) Development of phylogenetic markers for *Sebacina* (Sebacinaceae) mycorrhizal fungi associated with Australian orchids. *Applications in Plant Sciences* **2**(6), apps.1400015.

Ruibal MP, Peakall R, Smith LM, Linde CC. (2013) Phylogenetic and microsatellite markers for *Tulasnella* (Tulasnellaceae) mycorrhizal fungi associated with Australian orchids. *Applications in Plant Sciences* **1**(3).

Salazar GA, Cabrera LI, Madrinan S, Chase MW. (2009) Phylogenetic relationships of Cranichidinae and Prescottiinae (Orchidaceae, Cranichideae) inferred from plastid and nuclear DNA sequences. *Annals of Botany* **104**(3), 403–416.

Salazar GA, Van Den Berg C, Popovkin A. (2014) Phylogenetic relationships of *Discyphus scopulariae* (Orchidaceae, Cranichideae) inferred from plastid and nuclear DNA sequences: evidence supporting recognition of a new subtribe, Discyphinae. *Phytotaxa* **173**(2), 127–139.

Salzmann CC, Nardella AM, Cozzolino S, Schiestl FP. (2007) Variability in floral scent in rewarding and deceptive orchids: The signature of pollinator-imposed selection? *Annals of Botany* **100**(4), 757–765.

Scade A, Brundrett MC, Batty AL, Dixon KW, Sivasithamparam K. (2006) Survival of transplanted terrestrial orchid seedlings in urban bushland habitats with high or low weed cover. *Australian Journal of Botany* **54**(4), 383–389.

Schaffer GF, Peterson RL. (1993) Modifications to clearing methods used in combination with vital staining of roots colonized with vesicular-arbuscular mycorrhizal fungi. *Mycorrhiza* **4**, 29–35.

Schiestl F, Marion-Poll F. (2002) Detection of physiologically active flower volatiles using gas chromatography coupled with electroantennography. In *Molecular methods of plant analysis, Vol 21. Analysis of taste and aroma* (eds. JF Jackson, HF Linskens, and R Inman) pp. 173–198. Springer: Berlon, Germany.

Schiestl FP. (2005) On the success of a swindle: pollination by deception in orchids. *Naturwissenschaften* **92**(6), 255–264.

Schiestl FP, Ayasse M. (2002) Do changes in floral odor cause speciation in sexually deceptive orchids? *Plant Systematics and Evolution* **234**(1–4), 111–119.

Schiestl FP, Ayasse M, Paulus HF, Lofstedt C, Hansson BS, Ibarra F, Francke W. (2000) Sex pheromone mimicry in the early spider orchid (*Ophrys sphegodes*): patterns of hydrocarbons as the key mechanism for pollination by sexual deception. *Journal of Comparative Physiology a-Sensory Neural and Behavioral Physiology* **186**(6), 567–574.

Schiestl FP, Peakall R. (2005) Two orchids attract different pollinators with the same floral odour compound: ecological and evolutionary implications. *Functional Ecology* **19**(4), 674–680.

Schiestl FP, Peakall R, Mant JG, Ibarra F, Schulz C, Franke S, Francke W. (2003) The chemistry of sexual deception in an orchid-wasp pollination system. *Science* **302**(5644), 437–438.

Schiestl FP, Schluter PM. (2009) Floral isolation, specialized pollination, and pollinator behavior in orchids. *Annual Review of Entomology* **54**(1), 425–446.

Schodelbauerova I, Tremblay RL, Kindlmann P. (2010) Prediction vs. reality: can a PVA model predict population persistence 13 years later? *Biodiversity and Conservation* **19**(3), 637–650.

Schuiteman A, de Vogel E. (2003) Taxonomy for conservation. In *Orchid Conservation* (eds. KW Dixon, SP Kell, RL Barrett and PJ Cribb) pp. 55–68. Natural History Publications: Kota Kinabalu, Sabah, Malaysia.

Seaton PT, Hailes NSJ. (1989) Effect of temperature and moisture content on the viability of *Cattleya aurantiaca* seed. In *Modern Methods in Orchid Conservation* (ed. HW Pritchard) pp. 17–29. Cambridge University Press: Cambridge, UK.

Seaton PT, Hu H, Perner H, Pritchard HW. (2010) *Ex situ* conservation of orchids in a warming world. *The Botanical Review* **76**(2), 193–203.

Seaton PT, Pritchard HW. (2003) Orchid germplasm collection, storage and exchange. In *Orchid Conservation* (eds. KW Dixon, SP Kell, RL Barrett, and PJ Cribb) pp. 227–258. Natural History Publications: Kota Kinabalu, Sabah, Malaysia.

Seddon PJ. (2010) From reintroduction to assisted colonization: moving along the conservation translocation spectrum. *Restoration Ecology* **18**(6), 796–802.

Selosse M-A, Faccio A, Scappaticci G, Bonfante P. (2004) Chlorophyllous and achlorophyllous specimens of *Epipactis microphylla* (Neottieae, Orchidaceae) are associated with ectomycorrhizal septomycetes, including truffles. *Microbial Ecology* **47**(4), 416–426.

Shefferson RP, Kull T, Tali K. (2008) Mycorrhizal interactions of orchids colonizing Estonian mine tailings hills. *American Journal of Botany* **95**(2), 156–164.

Shefferson RP, Taylor DL, et al. (2007) The evolutionary history of mycorrhizal specificity among lady's slipper orchids. *Evolution* **61**(6), 1380–1390.

Shefferson RP, Weiss M, Kull T, Taylor DL. (2005) High specificity generally characterizes mycorrhizal association in rare lady's slipper orchids, genus *Cypripedium*. *Molecular Ecology* **14**, 613–626.

Simpson B, Neff J. (1983) *Evolution and diversity of floral rewards.* Van Nostrand Reinhold New York, 142–159.

Singer RB. (2002) The pollination mechanism in *Trigonidium obtusum* Lindl (Orchidaceae: Maxillariinae): sexual mimicry and trap-flowers. *Annals of Botany* **89**(2), 157–163.
Singh F. (1981) Differential staining of orchid seeds for viability testing. *American Orchid Society Bulletin* **50**, 416–418.
Smith D, Onions AHS. (1994) *The preservation and maintenance of living fungi*. CAB International: Commonwealth Mycological Institute, London, UK.
Smith SE, Read DJ. (1997) *Mycorrhizal symbiosis*. Academic Press: San Diego, U.S.
Smithson A, Gigord LDB. (2001) Are there fitness advantages in being a rewardless orchid? Reward supplementation experiments with *Barlia robertiana*. *Proceedings of the Royal Society B-Biological Sciences* **268**(1475), 1435–1441.
Smreciu EA, Currah RS. (1989) Symbiotic germination of seeds of terrestrial orchids of North America and Europe. *Lindleyana* **4**(1), 6–15.
Sneh B, Burpee L, Ogoshi A. (1991) Identification of *Rhizoctonia* species. APS Press: St. Paul, Minnesota, USA **133**(28).
Soliva M, Widmer A. (2003) Gene flow across species boundaries in sympatric, sexually deceptive *Ophrys* (Orchidaceae) species. *Evolution* **57**(10), 2252–2261.
Sommerville KD, Siemon JP, Wood CB, Offord CA. (2008) Simultaneous encapsulation of seed and mycorrhizal fungi for long-term storage and propagation of terrestrial orchids. *Australian Journal of Botany* **56**(7), 609–615.
Sonneveld C. (1979) Changes in chemical properties of soil caused by steam sterilization. In *Soil disinfestation* (ed. D Mulder) pp. 39–50. Elsevier: Amsterdam.
Sprengel CK. (1793) Das entdeckte Geheimnis der Natur im Bau und in der Befruchtung der Pflanzen. *Berlin: Friedrich Vieweg dem aeltern.*
Stamatakis A. (2006) RAxML-VI-HPC: maximum likelihood-based phylogenetic analyses with thousands of taxa and mixed models. *Bioinformatics* **22**(21), 2688–2690.
Stapleton JJ, Quick J, DeVay JE. (1985) Soil solarization: effects on soil properties, crop fertilisation and plant growth. *Soil Biology and Biochemistry* **17**, 369–373.
Steele WK. (1996) Large scale seedling production of North American *Cypripedium* species. In *North American Native Terrestrial Orchid Conference*, 1996, Germantown, Maryland, pp. 11–26.
Steiner KE, Whitehead VB, Johnson SD. (1994) Floral and pollinator divergence in two sexually deceptive South-African orchids. *American Journal of Botany* **81**(2), 185–194.
Stern WL, Warcup JH. (1994) Root tubercles in Apostasiad orchids. *American Journal of Botany* **81**(12), 1571–1575.
Stöckel M, Těšitelová T, Jersáková J, Bidartondo MI, Gebauer G. (2014) Carbon and nitrogen gain during the growth of orchid seedlings in nature. *New Phytologist* **202**(2), 606–615.
Stokl J, Paulus H, Dafni A, Schulz C, Francke W, Ayasse M. (2005) Pollinator attracting odour signals in sexually deceptive orchids of the *Ophrys fusca* group. *Plant Systematics and Evolution* **254**(1/2), 105–120.
Stokl J, Twele R, Erdmann DH, Francke W, Ayasse M. (2007) Comparison of the flower scent of the sexually deceptive orchid *Ophrys iricolor* and the female sex pheromone of its pollinator *Andrena morio*. *Chemoecology* **17**(4), 231–233.
Storfer A. (1996) Quantitative genetics: a promising approach for the assessment of genetic variation in endangered species. *Trends in Ecology & Evolution* **11**(8), 343–8.

Stoutamire WP. (1974) Terrestrial orchid seedlings. In *The orchids: scientific studies* (ed. CL Withner) pp. 101–128. John Wiley & Sons: New York, USA.

Sun M, Wong KC. (2001) Genetic structure of three orchid species with contrasting breeding systems using RAPD and allozyme markers. *American Journal of Botany* **88**(12), 2180–2188.

Swafford KL. (1999) *PAUP 4.0: phylogenetic analysis with parsimony (and other methods)*, version 4.0 b2a. Sinauer Associates Inc: Sunderland, Massachusetts, USA.

Swarts ND. (2007) Integrated conservation of the rare and endangered terrestrial orchid *Caladenia huegelii* HG Reichb (PhD thesis). PhD Thesis, University of Western Australia, Perth, Western Australia.

Swarts ND, Clements MA, Bower CC, Miller JT. (2014) Defining conservation units in a complex of morphologically similar, sexually deceptive, highly endangered orchids. *Biological Conservation* **174**, 55–64.

Swarts ND, Dixon KW. (2009) Terrestrial orchid conservation in the age of extinction. *Annals of Botany* **104**(3), 543–556.

Swarts ND, Sinclair EA, Francis A, Dixon KW. (2010) Ecological specialization in mycorrhizal symbiosis leads to rarity in an endangered orchid. *Molecular Ecology* **19**(15), 3226–3242.

Swarts ND, Sinclair EA, Krauss SL, Dixon KW. (2009) Genetic diversity in fragmented populations of the critically endangered spider orchid *Caladenia huegelii*: implications for conservation. *Conservation Genetics* **10**(5), 1199–1208.

Szalanski AL, Steinauer G, Bischof R, Petersen J. (2001) Origin and conservation genetics of the threatened Ute ladies'-tresses, *Spiranthes diluvialis* (Orchidaceae). *American Journal of Botany* **88**(1), 177.

Tali K, Fay MF, Bateman RM. (2006) Little genetic differentiation across Europe between early-flowering and late-flowering populations of the rapidly declining orchid *Neotinea ustulata*. *Biological Journal of the Linnean Society* **87**(1), 13–25.

Tamura K, Peterson D, Peterson N, Stecher G, Nei M, Kumar S. (2011) MEGA5: molecular evolutionary genetics analysis using maximum likelihood, evolutionary distance, and maximum parsimony methods. *Molecular Biology and Evolution* **28**(10), 2731–9.

Tan X-M, Chen X-M, Wang C-L, Jin X-H, Cui J-L, Chen J, Guo S-X, Zhao L-F. (2012) Isolation and identification of endophytic fungi in roots of nine *Holcoglossum* plants (Orchidaceae) collected from Yunnan, Guangxi, and Hainan provinces of China. *Current microbiology* **64**(2), 140–147.

Taylor DL, Bruns TD. (1999) Population, habitat and genetic correlates of mycorrhizal specialization in the 'cheating' orchids *Corallorhiza maculara* and *C. mertensiana*. *Molecular Ecology* **8**, 1719–1732.

Taylor DL, Bruns TD, Leake JR, Read DJ. (2002) Mycotrophic specialisation and function in myco-heterotrophic plants. In *Mycorrhizal ecology* (eds. MGA van der Heijden and I Sanders) pp. 375–413. Springer-Verlag: Berlin, Germany.

Taylor DL, Bruns TD, Szaro TM, Hodges SA. (2003) Divergence in mycorrhizal specialization within *Hexalectris spicata* (Orchidaceae), a nonphotosynthetic desert orchid. *American Journal of Botany* **90**(8), 1168–1179.

Taylor DL, McCormick MK. (2008) Internal transcribed spacer primers and sequences for improved characterization of basidiomycetous orchid mycorrhizas. *New Phytologist* **177**(4), 1020–1033.

Těšitelová T, Jersáková J, Roy M, Kubátová B, Těšitel J, Urfus T, Trávníček P, Suda J. (2013) Ploidy-specific symbiotic interactions: divergence of mycorrhizal fungi between cytotypes of the *Gymnadenia conopsea* group (Orchidaceae). *New Phytologist* **199**(4), 1022–1033.

Thompson KA, Newmaster SG. (2014) Molecular taxonomic tools provide more accurate estimates of species richness at less cost than traditional morphology-based taxonomic practices in a vegetation survey. *Biodiversity and conservation* **23**(6), 1411–1424.

Thompson PA. (1996) *Orchids from seed.* Royal Botanic Gardens: Kew, UK.

Threatened Species Section (2006). *Flora Recovery Plan: Tasmanian Threatened Orchids 2006-2010.* Department of Primary Industries, Water and Environment, Hobart.

Tisserant B, Gianinazzi-Pearson S, Gollotte A. (1993) *In planta* histochemical staining of fungal alkaline phosphatase activity for analysis of efficient arbuscular mycorrhizal infections. *Mycological Research* **97**, 245–250.

Touchell DH, Dixon KW. (1993) Cryopreservation of seed of Western Australian native species. *Biodiversity and Conservation* **2**, 594–602.

Tremblay RL. (1994) Frequency and consequences of multi-parental pollinations in a population of *Cypripedium calceolus* var. *pubescens* (orchidaceae). *Lindleyana* **9**(3), 161–167.

Tremblay RL, Ackerman JD, Zimmerman JK, Calvo RN. (2005) Variation in sexual reproduction in orchids and its evolutionary consequences: a spasmodic journey to diversification. *Biological Journal of the Linnean Society* **84**(1), 1–54.

Tremblay RL, Perez M, Larcombe M, Brown A, Quarmby J, Bickerton D, French G, Bould A. (2009) Population dynamics of *Caladenia*: Bayesian estimates of transition and extinction probabilities. *Australian Journal of Botany* **57**(4), 351–360.

Tremblay RL, Perez ME, Larcombe M, Brown A, Quarmby J, Bickerton D, French G, Bould A. (2009) Dormancy in *Caladenia*: a Bayesian approach to evaluating latency. *Australian Journal of Botany* **57**(4), 340–350.

Trinci APJ. (1971) Influence of the width of the peripheral growth zone on the radial growth rate of fungal colonies on solid media. *Microbiology* **67**(3), 325–344.

Trudell SA, Rygiewicz PT, Edmonds RL. (2003) Nitrogen and carbon stable isotope abundances support the myco-heterotrophic nature and host-specificity of certain achlorophyllous plants. *New Phytologist* **160**(2), 391–401.

Tsutsui K, Tomita M. (1989) Effect of plant density on the growth of seedlings of *Spiranthes sinensis* Ames and *Liparis nervosa* Lindl. in symbiotic culture. *Journal of the Japanese Society for Horticultural Science* **57**(4), 668–673.

Turnbull LA, Crawley MJ, Rees M. (2000) Are plant populations seed-limited? A review of seed sowing experiments. *Oikos* **88**(2), 225–238.

Van Waes JM, Debergh PC. (1986) Adaptation of the tetrazolium method for testing the seed viability, and scanning electron microscopy study of some Western European orchids. *Physiologia Plantarum* **66**, 435–442.

Vereecken NJ, Schiestl FP. (2008) The evolution of imperfect floral mimicry. *Proceedings of the National Academy of Sciences of the United States of America* **105**(21), 7484–7488.

Vitt P. (1991) Conservation of *Isotria medeoloides*: a federally endangered terrestrial orchid (Masters thesis). University of Maine.

Vitt P, Campbell CS. (1997) Reproductive biology of *Isotria medeoloides* (Orchidaceae). *Rhodora* **99**(897), 56–63.

Wallace L. (2002) Examining the effects of fragmentation on genetic variation in *Platanthera leucophaea* (Orchidaceae): inferences from allozyme and random amplified polymorphic DNA markers. *Plant Species Biology* **17**(1), 37–49.

Wallace LE. (2003) The cost of inbreeding in *Platanthera leucophaea* (Orchidaceae). *American Journal of Botany* **90**(2), 235–242.

Warcup JH. (1971) Specificity of mycorrhizal association in some Australian terrestrial orchids. *New Phytologist* **70**, 41–46.

Warcup JH. (1973) Symbiotic germination of some Australian terrestrial orchids. *New Phytologist* **72**, 387–392.

Warcup JH. (1975) Factors affecting symbiotic germination of orchid seed. In *Endomycorrhizas* (eds. FE Sanders, B Mosse, and PB Tinker) pp. 87–104. Academic Press: London, UK.

Warcup JH. (1981) The mycorrhizal relationships of Australian orchids. *New Phytologist* **87**, 371–381.

Warcup JH. (1985) *Rhizanthella gardneri* (Orchidaceae), it's *Rhizoctonia* endophyte and close association with *Melaleuca uncinata* (Myrtaceae) in Western Australia. *New Phytologist* **99**, 273–280.

Warcup JH, Talbot PHB. (1967) Perfect states of Rhizoctonias associated with orchids. I. *New Phytologist* **66**, 631–641.

Warcup JH, Talbot PHB. (1971) Perfect states of Rhizoctonias associated with orchids. II. *New Phytologist* **70**, 35–40.

Warcup JH, Talbot PHB. (1980) Perfect states of Rhizoctonias associated with orchids. III. *New Phytologist* **86**, 267–272.

Waud M, Busschaert P, Ruyters S, Jacquemyn H, Lievens B. (2014) Impact of primer choice on characterization of orchid mycorrhizal communities using 454 pyrosequencing. *Molecular Ecology Resources* **14**(4), 679–699.

Way MW. (2003) Chapter 9. Collecting seed from non-domesticated plants for long-term conservation. In *Seed Conservation: turning science into practice* (ed. RD Smith) pp. 163–201. Royal Botanic Gardens, Kew: UK.

Weller JI. (2009) *Quantitative trait loci analysis in animals.* CABI Publishing: London, UK.

Whigham DF. (1990) The effect of experimental defoliation on the growth and reproduction of a woodland orchid, *Tipularia discolor*. *Canadian Journal of Botany* **68**(8), 1812–1816.

White TJ, Bruns TD, Taylor LS. (1990) Amplification and direct sequencing of fungal ribosomal RNA genes for phylogenetics. In *PCR Protocols. A guide to method and applications* (eds. MA McInnis, DH Gelfand, JJ Sninsky, and TJ White) pp. 315–322. Academic Press: San Diego, California, USA.

Widmer A, Cozzolino S, Pellegrino G, Soliva M, Dafni A. (2000) Molecular analysis of orchid pollinaria and pollinaria-remains found on Insects. *Molecular Ecology* **9**, 1911–1914.

Wilcock C, Neiland M. (2002) Pollination failure in plants: why it happens and when it matters. *Trends in Plant Science* **7**, 270–277.

Willerslev E, Davison J, et al. (2014) Fifty thousand years of Arctic vegetation and megafaunal diet. *Nature* **506**(7486), 47–51.

Wong KC, Sun M. (1999) Reproductive biology and conservation genetics of *Goodyera procera* (Orchidaceae). *American Journal of Botany* **86**(10), 1406–1413.

Wood CB, Pritchard HW, Miller AP. (2000) Simultaneous preservation of orchid seed and its fungal symbiont using encapsulation-dehydration is dependent on moisture content and storage temperature. *Cryo-Letters* **21**(2), 125–136.
Worley AC, Sawich L, Ghazvini H, Ford BA. (2009) Hybridization and introgression between a rare and a common lady's slipper orchid, *Cypripedium candidum* and *C.parviflorum* (Orchidaceae). *Botany-Botanique* **87**(11), 1054–1065.
Wright S. (1965) The interpretation of population structure by F-statistics with special regard to systems of mating. *Evolution* **19**, 395–420.
Wunderlin R, Hansen B. (2004) Atlas of Florida vascular plants. In *Application development, Florida Center for Community Design and Research* (eds. S Landry and K Campbell): Institute for Systematic Botany, University of South Florida, Tampa, FL.
Wyatt R, Broyles SB. (1994) Ecology and evolution of reproduction in milkweeds. *Annual Review of Ecology and Systematics* **25**, 423–441.
Xing XK, Ma XT, Deng ZH, Chen J, Wu FZ, Guo SX. (2013) Specificity and preference of mycorrhizal associations in two species of the genus *Dendrobium* (Orchidaceae). *Mycorrhiza* **23**(4), 317–324.
Xu JT, Guo S. (2000) Retrospect on the research of the cultivation of *Gastrodia elata* Bl, a rare traditional Chinese medicine. *Chinese Medical Journal* **113**(8), 686–692.
Xu JT, Mu C. (1990) The relation between growth of *Gastrodia elata* protocorms and fungi. *Acta Botanica Sinica* **32**, 26–31.
Yagame T, Orihara T, Selosse MA, Yamato M, Iwase K. (2012) Mixotrophy of *Platanthera minor*, an orchid associated with ectomycorrhiza-forming Ceratobasidiaceae fungi. *New Phytologist* **193**(1), 178–187.
Yam TW, Nair H, Hew CS, Arditti J. (2002) Orchid seeds and their germination: an historical account. In *Orchid biology: reviews and perspectives*, VIII (eds. T Kull and J Arditti) pp. 387–504. Kluwer Academic Publishers Dordrecht.
Yamato M, Yagame T, Suzuki A, Iwase K. (2005) Isolation and identification of mycorrhizal fungi associating with an achlorophyllous plant, *Epipogium roseum* (Orchidaceae). *Mycoscience* **46**(2), 73–77.
Yang Q, Fu Y, Wang YQ, Wang Y, Zhang WH, Li XY, Reng YQ, Zhang J. (2014) Genetic diversity and differentiation in the critically endangered orchid (*Amitostigma hemipilioides*): implications for conservation. *Plant Systematics and Evolution* **300**(5), 871–879.
Yeh F, Yang R, Boyle T, Ye ZH, Mao J. (1997) POPGENE, the user-friendly shareware for population genetics analysis, 1.21. Molecular Biology and Biotechnology Centre, University of Alberta, Edmonton, Canada.
Yuan L, Yang ZL, Li SY, Hu H, Huang JL. (2010) Mycorrhizal specificity, preference, and plasticity of six slipper orchids from South Western China. *Mycorrhiza* **20**(8), 559–568.
Zelmer CD, Cuthbertson L, Currah RS. (1996) Fungi associated with terrestrial orchid mycorrhizas, seeds and protocorms. *Mycoscience* **37**(4), 439–448.
Zettler LW, Barrington FV, McInnis TM, Jr. (1995) Development morphology of *Spiranthes odorata* seedlings in symbiotic culture. *Lindleyana* **10**(3), 211–216.
Zettler LW, Corey LL, Jacks AL, Gruender LT, Lopez AM. (2013) *Tulasnella irregularis* (Basidiomycota: Tulasnellaceae) from roots of *Encyclia tampensis* in south Florida, and confirmation of its mycorrhizal significance through symbiotic seed germination. *Lankesteriana* **13**(1–2), 119–128.

Zettler LW, David SE. (2014) Endangered Hawaiian endemic, *Peristylus holochila*, initiates anthesis *in vitro* and *ex vitro*. *Orchid Conservation News* (May), 1–4.

Zettler LW, Hofer CJ. (1998) Propagation of the little club-spur orchid (*Platanthera clavellata*) by symbiotic seed germination and its ecological implications. *Environmental and Experimental Botany* **39**(3), 189–195.

Zettler LW, Mcinnis JTH. (1992) Propagation of *Platanthera integrilabia* (Correll) Luer, an endangered terrestrial orchid, through symbiotic seed germination. *Lindleyana* **7**(3), 154–161.

Zettler LW, Oppenheimer H. (2012) Hawaii's native orchids: a closer look at the trio. *Orchids—the Bulletin of the American Orchid Society* **81**(2), 100–101.

Zettler LW, Perlman S, Dennis DJ, Hopkins SE, Poulter SB. (2005) Symbiotic germination of a federally endangered Hawaiian endemic, *Platanthera holochila* (Orchidaceae), using a mycobiont from Florida: a conservation dilemma. *Selbyana*, 269–276.

Zettler LW, Perlman SP. (2012) *Peristylus holochila*—Hawaii's rarest native orchid and the battle to save it from extinction. *Orchids—the Bulletin of the American Orchid Society* **81**(2), 94–99.

Zettler LW, Sharma J, Rasmussen FN. (2003) Mycorrhizal diversity. In *Orchid Conservation* (eds. KW Dixon, SP Kell, RL Barrett, and PJ Cribb) pp. 205–226. Natural History Publications Kota Kinabalu, Sabah, Malaysia.

Zettler LW, Wood EM, Johnson LJ, Kirk AK, Perlman SP. (2011) Seed propagation and reintroduction of the U.S. federally endangered Hawaiian endemic, *Platanthera holochila* (Hbd.) Krzl. (Orchidaceae). *European Journal of Environmental Sciences* **1**(2).

Zhai JW, Zhang GQ, et al. (2013) A new orchid genus, *Danxiaorchis*, and phylogenetic analysis of the tribe Calypsoeae. *Plos One* **8**(4).

Zhang Y, Lee Y-I, Deng L, Zhao S. (2013) Asymbiotic germination of immature seeds and the seedling development of *Cypripedium macranthos* Sw., an endangered lady's slipper orchid. *Scientia Horticulturae* **164**, 130–136.

Zi X-M, Sheng C-L, Goodale UM, Shao S-C, Gao J-Y. (2014) *In situ* seed baiting to isolate germination-enhancing fungi for an epiphytic orchid, *Dendrobium aphyllum* (Orchidaceae). *Mycorrhiza* **24**(7), 487–499.

Ziegler H. (1995) Deuterium content in organic material of hosts and their parasites. In *Ecophysiology of photosynthesis* (eds. ED Schulze and MM Caldwell) pp. 393–408. Springer.

Zimmerman JK. (1980) Ecological correlates of labile sex expression in the orchid *Catasetum viridiflavum*. *Ecology* **72**, 597–608.

INDEX

Note: Page numbers followed by "f" indicate a figure and by "t" indicate a table.